SERIES 3—NINTH EDITION

Technical Drawing Problems

Henry Cecil Spencer
Late Professor Emeritus of Technical Drawing;
Formerly Director of Department
Illinois Institute of Technology

Ivan Leroy Hill
Professor Emeritus of Engineering Graphics;
Formerly Chairman of Department
Illinois Institute of Technology

John Thomas Dygdon
Professor of Engineering Graphics,
Chairman of the Department,
and Director of the Division of Academic Services
and Office of Educational Services
Illinois Institute of Technology

James E. Novak
Associate Director/Executive Officer,
Office of Educational Services
Illinois Institute of Technology

Macmillan Publishing Company
New York
Maxwell Macmillan Canada
Toronto
Maxwell Macmillan International
New York   Oxford   Singapore   Sydney

Copyright © 1991 by Macmillan Publishing Company.

Laser typesetting by **Ewing Systems**, 409 W. 24th St., Suite #14, New York, NY 10011. Printed in the United States of America.

All rights reserved. No part of this book may be reproduced or transmitted in any form or by any means, electronic or mechanical, including photocopying, recording, or any information storage and retrieval system, without permission in writing from the Publisher.

Earlier editions copyright © 1960 and copyright © 1974 and 1980 by Macmillan Publishing Company.

Macmillan Publishing Company
866 Third Avenue, New York, New York 10022

Macmillan Publishing Company is part of the Maxwell Communication Group of Companies.

Maxwell Macmillan Canada, Inc.
1200 Eglinton Avenue East
Suite 200
Don Mills, Ontario M3C 3N1

ISBN 0-02-414630-7

Printing: 1 2 3 4 5 6 7 8   Year: 1 2 3 4 5 6 7 8 9 0

Preface

Technical Drawing Problems, Series 3, is intended primarily for use with *Technical Drawing* (Ninth Edition, 1991) by Giesecke, Mitchell, Spencer, Hill, Dygdon, and Novak, and all references and instructions refer to that text. However, this workbook may be used with any good reference text.

Since the time available for technical drawing has become increasingly limited, the objective has been to produce in as few sheets as possible a complete coverage of the basic fundamentals. A number of problems are printed on vellum to provide experience in the manner of commercial practice. These are in a group following drawing number 88. It is expected that in most cases the instructor will supplement these problem sheets with assignments of problems from the text to be drawn on blank paper.

All of the problems are based upon actual industrial designs, and their presentations are in accord with the new American National Standard Drafting Manual, Y14, and other relevant ANSI standards.

An outstanding feature of this revision is the minimum use of fractional-inch dimensions and the emphasis on decimal-inch and metric dimensions now used extensively in industry. A decimal and millimeter equivalents table and appropriate full-size and half-size scales are provided inside the front and back covers for the student's convenience.

Less space is devoted to lettering and other exercises intended primarily to develop skills, while much more emphasis is given to technical sketching. Since the engineer has at his disposal all sorts of cross-section papers to assist in sketching, numerous problems in this workbook make use of similar grids. In addition, many of the instrumental drawing problems are equally suitable for freehand sketching.

A special effort has been made to present problems that are thought-provoking rather than requiring a great deal of routine drafting.

In response to the increased usage of computer technology for drafting and design, a number of problem sheets in computer-aided drafting (CAD) have been included. The problem sheets on detail drawings are presented to provide practice in making regular working drawings of the type used in industry. These are suggested for solution either by a computer-aided drafting system or by traditional drafting methods.

All sheets are 8.5" × 11.0" in conformity with the American National Drafting Standards. This size also facilitates handling and filing by the instructor. In general a 4H lead will be found suitable for construction lines and guide lines for lettering; a 2H lead for center lines, section lines, dimension lines, extension lines, and phantom lines; and an F lead for general line work and lettering. The student should make all construction lines very light; *they should not be erased.*

The Instructions provide detailed information together with references to the text for each problem. The student is urged to study these Instructions and references carefully before starting each problem.

The many valuable suggestions from colleagues who use this material are sincerely appreciated. The cooperation of industrial firms who so generously supplied problem material is gratefully acknowledged. Additional comments and criticisms from users of this workbook will be most welcome.

Ivan Leroy Hill
Clearwater, FL

John Thomas Dygdon
*Illinois Institute
of Technology
Chicago, IL*

James E. Novak
*Illinois Institute
of Technology
Chicago, IL*

Contents

Instructions 1

Worksheets

Topics	*Drawings*
Vertical Lettering	1–4
Inclined Lettering	5–8
Mechanical Drawing	9–12
Geometric Constructions	13–15
Multiview and Isometric Sketching	16–20
Multiview Projection Problems	21–31
Sectioning	32–39
Auxiliary Views	40–48
Revolutions	49, 50
Isometric Drawing	51–56
Oblique Drawing	57–59
Dimensioning	60–67
Threads and Fasteners	68–72
Welding	73
Charts and Graphs	74, 75
Engineering Graphics	76–81
Computer-Aided Drafting	82–86
Detail Drawings	87, 88

Vellums

The following drawings, printed on vellum, appear at the back of this workbook, after sheet number 88.

Vertical Lettering	3
Inclined Lettering	7
Mechanical Drawing	10, 12
Multiview Projection Problems	28, 31
Sectioning	36, 38, 39
Auxiliary Views	45, 48
Isometric Drawing	56
Dimensioning	62, 65
Threads and Fasteners	71
Welding	73

Instructions

References are to Ninth Edition of *Technical Drawing (1991)*
by Giesecke, Mitchell, Spencer, Hill, Dygdon, and Novak

Throughout this workbook alternative dimensions, often not the *exact* equivalents, are given in millimeters and inches. Although it is understood that 25.4 mm = 1.00", it is more practical to use approximate equivalents such as 25 mm for 1.00"; 12.5 or 12 mm for .50"; 6 mm for .25"; 3 mm for .12"; etc. Exact equivalents should be used when accurate fit or critical strength is involved.

Drawing 1. Vertical Capitals and Numerals. References: §§4.1–4.18. Using an HB lead, letter the indicated characters in the spaces provided. These large letters and numerals may be sketched lightly first, and then corrected where necessary before being made heavy with the strokes shown. Omit the numbers and arrows from your letters. All lettering must be *clean-cut* and **black**. In the title strip, under **DRAWN BY**, draw light guide lines from the starting marks shown, plus random vertical guidelines, and letter your name with the last name first. Under **FILE NO.** letter identification symbol as assigned by your instructor.

Drawing 2. Vertical Capitals and Numerals. References: §§4.1–4.18, 4.24. First, draw light vertical guide lines at random from bottom to top of the sheet. Do not draw separate vertical guide lines for each line of lettering. Reproduce the lettering as exactly as you can, using an HB lead for the larger letters and a sharp F lead for the smaller letters. Note that the last line of lettering is to be lettered twice. All letters must be *clean-cut* and **black**.

Drawing 3. Vertical Capitals and Numerals. References: §§4.10, 4.14–4.18, 4.20, 4.24, 13.5, 13.7–13.11, 13.13–13.15, 13.17. On the left side of the sheet are shown a number of lettering applications. On the right, reproduce the lettering, arrowheads, and finish marks, using a sharp F lead. Except for the title **TOOL HOLDER** at the bottom, all lettering on the sheet is 3 mm or .12" high. Draw all guide lines with the aid of a lettering triangle or the Ames Lettering Guide, using a 4H or 6H lead. Letter, in vertical capitals, the title **TOOL HOLDER**, etc., on center as shown in Fig.4.36 (b). Use height and spacing as specified. Underline **TOOL HOLDER**. All lettering must be *clean-cut* and **black**.

Drawing 4. Vertical Lowercase Lettering. References: §§4.21, 4.22, 4.24. In the upper half of the sheet, fill in the letters in the spaces provided, using an HB or F lead. Use the strokes shown, but omit the numbers and arrows from your letters. In the lower half of the sheet, draw vertical guide lines and then letter each line of lettering twice on the guide lines provided, using a sharp F lead. Make all letters *clean-cut* and **black**.

Drawing 5. Inclined Capitals and Numerals. References: §§4.1–4.17, 4.19. Using an HB lead, letter the indicated characters in the spaces provided. These large letters and numerals may be sketched lightly first, and then corrected as necessary before being made heavy with the strokes shown. Omit the numbers and arrows from your letters. All lettering must be *clean-cut* and **black**. In the title strip, under **DRAWN BY**, draw light guide lines from the starting marks shown, and letter your name with the last name first. Under **FILE NO.** letter identification symbol as assigned by your instructor.

Drawing 6. Inclined Capitals and Numerals. References: §§4.1–4.17, 4.19, 4.24. First, draw light inclined guide lines at random from bottom to top of the sheet. Do not draw separate inclined guide lines for each line of lettering. Reproduce the lettering as exactly as you can, using an HB lead for the larger letters and a sharp F lead for the smaller letters. Note that the last line of lettering is to be lettered twice. All letters must be *clean-cut* and **black**.

Drawing 7. Inclined Capitals and Numerals. References: §§4.10, 4.14–4.17, 4.19, 4.20, 4.24, 13.5, 13.7–13.11, 13.13–13.15, 13.17. On the left side of the sheet are shown a number of lettering applications. On the right, reproduce the lettering, arrowheads, and finish marks, using a sharp F lead. Except for the title **TOOL HOLDER** at the bottom, all lettering on the sheet is 3 mm or .12" high. Draw all guide lines with the aid of a lettering triangle or the Ames Lettering Guide, using a 4H or 6H lead. Letter the title **TOOL HOLDER**, etc., on center as shown in Fig. 4.36 (b). All lettering must be *clean-cut* and **black**.

Drawing 8. Inclined Lowercase Lettering. References: §§4.21, 4.23, 4.24. In the upper half of the sheet, fill in the letters in the spaces provided, using an HB or F lead. Use the strokes shown, but omit the numbers and arrows from your letters. In the lower half of the sheet, draw inclined guide lines and then letter each line of lettering twice on the guide lines provided, using a sharp F lead. Make all letters *clean-cut* and **black**.

Drawing 9. Alphabet of Lines. References: §§2.1–2.17, 2.20–2.31.

Spaces 1 and 2. Draw indicated lines full length in the given spaces.

Space 3. Section lines at 30° with horizontal upward to the right and 12.5 mm (.50") apart are required. First find center of space by drawing diagonals in *very light* construction lines. Through center draw construction line 60° with horizontal upward to the left and, beginning at the center, set off 12.5 mm (.50") intervals. Draw required section lines through these points to fill space.

Space 4. Using similar construction methods to those outlined for *Space 3,* draw visible lines at 75° with horizontal upward to the left at 12.5 mm (.50") intervals.

Space 5. Draw center lines parallel to given line and at 12.5 mm (.50") intervals to fill the space.

Space 6. Draw alternate visible and hidden lines 12.5 mm (.50") apart and perpendicular to the given line. Arrange so that one visible line passes through center of space.

Drawing 10. Scales and Layout. References: §§2.18, 2.24–2.43, 2.46, 2.47, 17.1–17.6.

Space 1. Use architects, engineers or metric scales as necessary. Measure lines A, C, and F at the scales shown, and indicate the scaled lengths (L) at the right. At B, D, E, G, and H draw lines of specified lengths at scales shown. Terminate the lines in the same manner as for given lines. At J through L, determine the scales and lengths of lines and record the scales and lengths in the spaces provided.

Line J is over 500' and under 600' in length.
Line K is between 530 m and 550 m in length.
Line L is one twenty-fourth size.

Space 2. Draw the two views of the Anchor Bracket full size, locating the views by the starting corners indicated. Your final lines should approximate those shown in Fig. 2.14, with distinct differences in line thicknesses.

Drawing 11. Layout, Chart, and Graph. References: §§2.25–2.27, 2.33–2.43, 2.46, 2.53, 2.54, 28.1–28.4, 28.16.

Space 1. Draw the two views of the Index Arm full size, locating the views by the starting points indicated. Use engineers scale.

Space 2. Draw bar chart representing the following U.S. resident population age structure for 1970. Use 6 mm (.25") wide bars, and cross-hatch with 45° section lines approximately 3 mm or .12" apart. Draw the horizontal grid lines from the main divisions, but do not cross the bars.

Age Group	Millions	Age Group	Millions
Under 10	37.6	40 to 49	24.2
10 to 19	39.8	50 to 59	20.8
20 to 29	29.4	60 to 69	15.6
30 to 39	22.6	Over 70	13.0

Space 3. Plot the following data regarding cutting speeds and machining costs, after establishing the grid with construction lines. Indicate the plotted points by small circles approximately 1.5 mm (.06") dia. drawn with a template or the bow pencil. Draw a smooth visible line through the plotted points, and complete the grid with fine **black** lines. Do not draw through the small circles.

Cutting Speed, ft/min	Cost, $/piece	Cutting Speed, ft/min	Cost $/piece
145	.26	265	.08
150	.24	300	.068
157.5	.20	400	.042
175	.16	500	.032
205	.12	600	.028

Drawing 12. Layout with Dimensions. References: §§2.2, 2.11, 2.13, 2.31–2.42, 2.45, 2.46, 7.4, 7.7, 7.33, 17.1–17.6. Draw the views of the Guide Bracket to the scale indicated, spacing the views according to the given locating points. Include all dimensions, notes, and finish marks. Use 3 mm or .12" high lettering, vertical or inclined as assigned. Space dimension lines 10 mm (.40") from the views and 10 mm (.40") apart. Make your lines *clean-cut* and **black** in conformity with the pencil lines in Fig. 2.16. Space the section lines about 3 mm or .12" apart. Spotface depths are usually not specified, but they are commonly drawn 1.5 mm (.06") deep.

Drawing 13. Geometric Constructions. References: §§5.1–5.6. Show light construction on all problems. Add center lines where necessary.

Space 1. References: §§5.8, 5.9. Locate and draw hole as indicated. Add center lines.

Space 2. Reference: 5.10. Complete the view as indicated.

Space 3. References: §§5.14, 5.15. Divide speedometer dial into 10 km/h intervals from 0 to 100. Show interval marks and numbering as shown in small drawing.

Space 4. Reference: §5.18. Locate centers for holes as specified. Draw holes and center lines.

Space 5. Reference: §5.19. Draw 90° notch as specified, and complete the view.

Space 6. Reference: Fig. 5.23 (c). Draw the square traffic sign to scale as shown. Omit lettering.

Drawing 14. Geometric Constructions. Show light construction on all problems.

Space 1. References: Fig. 5.26 (c) and (d). Complete the end view of the 12-point (double hexagon) socket as indicated.

Space 2. Reference: Fig. 5.28 (a). Draw the view of the octagonal face only of the engineers' cross-peen hammer head, which is cut from 28 mm square stock.

Instructions

Space 3. Reference: Fig. 5.32 (a). Determine diameter of milling cutter to cut true arc through points A, B, and C. Give diameter to nearest 0.5 mm.

Space 4. References: §§2.54, 5.51. Using the concentric-circle method, draw profile of 56 mm × 82 mm elliptical cam, starting at point A. Use a minimum of 16 points to establish the ellipse. Draw ellipse with the aid of the irregular curve.

Space 5. Reference: §5.57. Draw approximate ellipse as indicated.

Space 6. Reference: Fig. 5.57 (c) and (d). Draw parabola as indicated, and find focus.

Drawing 15. Tangencies. Show light construction on all problems, and indicate all tangent points with light lines as shown in the small view in *Space 1*.

Space 1. Reference: Fig. 5.35 (a). Complete the outline of the Gasket.

Space 2. References: §§5.37, 5.38. Complete the layout of the Slide Bracket.

Space 3. References: §§5.39, 5.40. Finish the half-view of the Cam.

Space 4. References: Figs. 5.36 (c), 5.41 (a). Complete the layout of the Pull Knob.

Space 5. Reference: Fig. 5.40 (a). Complete the view of the Grinder Rest.

Space 6. References: Figs. 5.40 (b), 5.41. Complete the outline of the Pivot Bracket.

Drawing 16. Multiview and Isometric Sketching. References: §§6.1–6.8, 6.12, 6.14, 6.18–6.22, 6.25, 6.26, 6.28–6.31.

Spaces 1 and 2. Sketch front, top, and right-side views of two objects assigned from A to D. Allow three squares between views.

Spaces 3 and 4. Sketch isometrics of two objects assigned from E to H.

Drawing 17. Multiview Technical Sketching. References: §§6.1–6.10, 6.18–6.31. Sketch the views as indicated. Make all final lines *clean-cut* and **black**. In the space provided, letter the names of the necessary views.

Drawing 18. Multiview and Isometric Technical Sketching. References: §§6.1–6.14, 6.18–6.31. Sketch the views as indicated, spacing the views three squares apart. Make all final lines *clean-cut* and **black** so the sketches will stand out from the grid lines.

Drawing 19. Choice of Views and Multiview Sketching. References: §§6.5–6.10, 6.18–6.31. At the top of the sheet are shown six objects with arrows indicating the front views. In each case indicate the correct choice of views by neatly checking (thus: ✓) in the table below each pictorial. In the lower half of the sheet, sketch the necessary views of one of the six problems, as assigned by your instructor. Make the views large enough to fit the space comfortably. It is most important to establish the views in correct proportion, as explained in §6.10.

Drawing 20. Multiview Technical Sketching. References: §§6.5–6.10, 6.18–6.31. Sketch views of Thread Roll Holder, as indicated. Good proportions are most important, §6.10. All holes are understood to be through holes. Note given corners for positioning the views.

Drawing 21. Missing Lines. References: §§6.5, 6.18–6.20, 6.25, 6.26, 6.28–6.31. Each problem is incomplete because lines are missing from one or more views. Add all missing lines freehand, including center lines.

Drawing 22. Missing Lines. References: §§6.25, 6.26, 7.1–7.14. Each problem is incomplete because lines are missing from one or more views. Add all missing lines, using instruments. Note that "missing lines" may include center lines.

Drawing 23. Missing Views. References: §§6.25, 6.26, 7.1–7.14. In each problem two complete views are given, and a third view is missing. Add the third view in each case freehand.

Drawing 24. Missing Views. References: §§6.25, 6.26, 7.1–7.30, 7.32, 7.33. In each problem two views are given. Add the third view, using instruments. The given views are complete except in Prob. 4 where lines are missing in the side view.

Drawing 25. Missing Views. References: §§6.25, 6.26, 7.1–7.30, 7.32, 7.33. In each problem two complete views are given. Add the third view in each case, using instruments.

Drawing 26. Missing Views. References: §§7.1–7.28, 7.32. In each problem two complete views are given. Add the third view in each case, using instruments.

Drawing 27. Missing Views. References: §§7.1–7.30, 7.33. In each problem two complete views are given. Add the third view in each case, using instruments. For Prob.1, study Fig. 7.35.

Drawing 28. Missing Views. References: §§7.1–7.30, 7.33. In each problem two complete views are given. Add the third view in each case, using instruments.

Drawing 29. Missing Views. References: §§7.1–7.30, 7.33, 7.34, 13.17. In each problem two complete views are given. Add the third view in each case, using instruments. Add all finish marks.

Drawing 30. Missing View. References: §§7.1–7.29, 7.34, 7.35. Two complete views are given. Add the top view, using instruments.

Drawing 31. Missing View. References: §§7.1–7.29, 7.34, 7.35, 13.17. Two complete views are given. Add the right-side view, using instruments. Show all finish marks. Draw small fillets and rounds freehand.

Drawing 32. Sectional Views. References: §§9.1–9.7, 9.15. Sketch the sections as indicated, using an HB or F lead for visible lines and a sharp F lead for section lines and center lines. Make sections lines thin to contrast well with heavy visible lines. All lines should be *clean-cut* and **black**. No additional cutting planes are required.

Drawing 33. Sectional Views. References: §§9.1–9.7, 9.12. Draw the indicated sectional views, using instruments. Add finish marks in Prob. 2. See §13.17. In Prob. 5, do not revolve lug to produce an aligned section. See §9.13. In Prob. 6, show hidden lines in unsectioned half of the half-section view.

Drawing 34. Sectional Views. References: §§2.11, 9.5, 9.8, 9.9, 9.11–9.13, 9.15. Draw the indicated sectional views, using instruments. In all four problems, fillets and rounds are 1.5 mm R, to be drawn freehand. In Prob. 2, draw revolved section in the opening in the top view, and show break lines on each side of the section. See Fig. 9.18. Draw break line from A to B in the front view; then draw broken-out section to the right of the break line.

Drawing 35. Sectional Views. References: §§9.1–9.6, 9.12, 9.15. Draw the indicated sectional views, using instruments. In Prob. 3, all fillets and rounds are 1.5 mm R, to be drawn freehand. In this problem add all finish marks (except holes).

Drawing 36. Sectional Views. References: §§9.1–9.6, 16.21. Draw the indicated sectional views, using instruments. In Prob. 2 is shown a portion of an assembly full section with a round shaft extending through a cast-iron cover and a steel plate, which are held together by bolts. Section-line the sectioned areas, using symbolic section lining for each material.

Drawing 37. Sectional Views. References: §§7.34, 9.1–9.6, 9.10, 12.5, 13.17. Draw the indicated sectional views, using instruments, In Prob. 1, add finish marks (except for holes).

In Prob. 2, show all visible lines behind the cutting plane in each section.

Drawing 38. Sectional View. References: §§7.33, 7.34, 9.13, 9.15, 12.5. Draw Section A–A as indicated.

Drawing 39. Sectional Views. References: §§7.34, 9.1–9.6, 12.5. Draw indicated sections.

Drawing 40. Primary Auxiliary Views. References: §§10.1–10.10. Sketch the auxiliary views as indicated. Using an HB or F lead, make visible lines and hidden lines **black** so that the views will stand out clearly from the grids. Use a sharp F lead for the center lines. Letter folding lines as shown in Fig. 10.3, and reference planes as in Fig. 10.6. In Probs. 3 and 4, include all hidden lines.

Drawing 41. Primary Auxiliary Views. References: §§10.1–10.10, 10.14, 10.16, 10.17. Using instruments, add any missing lines in the regular views or auxiliary views.

Drawing 42. Primary Auxiliary Views. References: §§10.1–10.14, 10.16, 13.14. Using instruments, draw the indicated views. Use folding lines in Prob. 1 and reference-plane lines in Probs. 2, 3, and 4. In Prob. 3, dimension the angles between surfaces A and B, and A and C.

Drawing 43. Primary Auxiliary Views. References: §§10.1–10.11, 10.14, 10.17, 13.14. Using instruments, draw the indicated auxiliary views. In Probs. 1 and 2, dimension the required angles in degrees. See Fig. 10.22. In Prob. 3, make the numbers the same size as those given (1.5 mm or .06" high). Use reference-plane lines or folding lines.

Drawing 44. Primary Auxiliary Views. References: §§10.1–10.11. Using instruments, draw the indicated views. Dimension the required views. Dimension the required angles in degrees. See Fig. 13.17. Use folding lines or reference-plane lines.

Drawing 45. Primary Auxiliary Views. References: §§10.1–10.11, 10.14. Using instruments, draw the indicated auxiliary views. Use reference-plane lines or folding lines.

Drawing 46. Secondary Auxiliary Views. References: §§10.19–10.22. Using instruments, draw the indicated auxiliary views.
Space 1. Surface A is a normal surface, §7.19, and therefore appears true size in the given view. Use reference-plane lines. Dimension the 135° angle, Fig. 13.17.
Space 2. The height of the object is shown in the reduced-scale drawing. Draw the primary auxiliary view 22 mm from the given top view. Use folding lines. Dimension the angle between surface A and B in degrees, Fig. 13.17.

Drawing 47. Secondary Auxiliary Views. References: §§10.19–10.24. Using instruments, draw or complete the indicated views. Use folding lines or reference-plane lines in both problems. Show all hidden lines.

Drawing 48. Secondary Auxiliary Views. References: §§10.19, 10.21, 10.24. Using instruments, draw the indicated views. Use folding-line method. Space the views about 25 mm apart. Number at least eight corners, making numerals 1.5 mm (.06") high. Dimension the angle between surfaces A and B in degrees, Fig. 13.17.

Drawing 49. Primary Revolutions. References: §§11.1–11.13. Using instruments, draw the indicated views. Locate the revolved views by the starting points P.

Drawing 50. Primary Revolutions. References: §§7.10, 11.1–11.4, 11.11. Using instruments, draw the indicated constructions. In Probs. 1 and 2, use alternate position lines to show the revolved surfaces, as in Fig. 11.11.

Drawing 51. Isometric Sketching. References: §§6.12–6.14. Make isometric sketches of objects shown, using the starting corners indicated. Omit hidden lines and center lines. Make visible lines **black** so isometrics will stand out clearly from the grids.

Drawing 52. Isometric Drawing. References: §§18.3–18.15, 18.18, 18.20. Using instruments, make isometric drawings of two assigned problems. Use the indicated starting corners. Show all construction.

Drawing 53. Isometric Drawing. References: §§18.13, 18.18, 18.20, 18.24. Using instruments, make isometric drawings. Use dividers to transfer all measurements from the

Instructions

views to the isometrics. Double each measurement to produce isometrics to full scale.

Drawing 54. Isometric Drawing. References: §§18.13, 18.18, 18.20, 18.24. Using instruments, make isometric drawings. Use dividers to transfer all measurements from the views to the isometrics. Double each measurement to produce isometrics to full scale.

Drawing 55. Isometric Drawing. References: §§18.13–18.18. Using instruments, make isometric drawings. In Prob. 2, use dividers to double distances and to transfer them from the given views to the isometric.

Drawing 56. Isometric Drawing. References: §§18.15, 18.16, 18.18, 18.20. Using instruments, make isometric drawing as specified. Use dividers to double distances and to transfer them from the given views to the isometric. For the curved arm, use the method shown in Fig. 18.20, but construct the sections horizontally.

Drawing 57. Oblique Sketching. References: §§6.3, 6.15, 6.16. Using the method shown in Fig. 6.28, make oblique sketches of objects shown, using the starting corners indicated. Omit hidden lines and center lines. Make visible lines **black** so sketches will stand out clearly from the grids.

Drawing 58. Oblique Projection. References: §§19.1–19.6. Draw cavalier or cabinet projections as indicated and as assigned, using instruments. Show construction for all points of tangency. Omit hidden lines and center lines.

Drawing 59. Oblique Projection. References: §§19.1–19.6, 19.10. Draw cavalier or cabinet projections as assigned, using instruments. Show construction for all points of tangency. Omit hidden lines and center lines. Omit finish marks unless assigned.

Drawing 60. Dimensioning. References: §§13.1–13.25, 13.30, 13.31. Use the complete decimal dimensioning system with metric values. If assigned, use decimal-inch equivalents. Add dimensions freehand, spacing dimension lines approximately 10 mm from the views and 10 mm apart. Include necessary finish marks in Prob. 2. Note that in Prob. 2 the drawing is half the size of the actual part. Dimension the keyway in the manner shown in Fig. 13.44 (x). The two small holes are drilled, Appendix 11, and the large hole is bored. In the bored-hole note, specify the diameter to two places.

Drawing 61. Dimensioning. References: §§13.4–13.25, 13.30, 13.31. Use the complete decimal dimensioning system with metric values. If assigned, use decimal-inch equivalents. Add dimensions mechanically, spacing dimension lines 10 mm from the views and 10 mm apart.

Space 1. Since the CRS (cold-rolled steel), the object is understood to be finished all over, and no finish marks are necessary. In this problem the small hole is drilled, and the two medium-sized holes are drilled and countersunk. See Fig.13.44 (c) and Appendix 11. The large hole is reamed. In the note indicate this diameter in decimals to two places.

Space 2. Add finish marks. The small holes are drilled and equally spaced. The central portion is cored, and the large end holes are bored. Use the note **19.05–19.10 BORE**. All fillets and rounds are 3R.

Drawing 62. Dimensioning. References: §§13.4–13.25, 13.30–13.31. Use the complete decimal dimensioning system with metric values. If assigned, use decimal-inch equivalents. Add dimensions mechanically, spacing dimension lines 10 mm from the views and 10 mm apart. The three large holes (two in line) are reamed. Use the note **15.82–15.88 REAM, 3 HOLES**. The small holes are drilled. Since the material is CRS (cold-rolled steel), the object is understood to be finished all over, and no finish marks are necessary.

Drawing 63. Mating Parts Dimensioning. References: §§13.9, 13.16, 13.17, 13.20–13.27, 13.31, 15.21. The T-Slot Clamp is a holding device used in the machine shop to hold work pieces in position. Dimension the parts completely, spacing dimension lines 10 mm from the views and 10 mm apart. Use the complete decimal dimensioning system with metric values. If assigned, use decimal-inch equivalents. Note that finish marks are not required on parts made of CRS (cold-rolled steel).

Part No. 1. Frame. The hole in the base is drilled 1.5 mm (.06") larger than the M10 × 1.5 T-Slot Bolt. The bottom of the base is machined flat. For the tapped hole use metric coarse threads or if assigned, use Unified Coarse threads.

Drawing 64. Mating Parts Dimensioning (Continued). References: Same as for Drawing 63.

Part No. 2. Clamp Screw. The part is cylindrical except for the spherical ball on the end. The threads correspond to those in the Frame. The length of the threaded portion, including the chamfer and relief, is 66 mm. The drilled hole is 0.6 mm larger than the nominal size of the Handle.

Part No. 3. Pad. The drilled hole is 0.5 mm larger than the ball of the Clamp Screw. The Pad is attached to the ball of the Clamp Screw by crimping the slit edges.

Part No. 4. Handle. The Handle is 90 mm long. The diameter is 6.375–6.385 mm.

Part No. 5. Handle Cap. The Cap is cylindrical. The hole is drilled and reamed to 6.350–6.365 mm diameter so as to fit tightly on the Handle.

Drawing 65. Tolerance Dimensioning. References: §§13.9, 13.20–13.25, 13.27, 13.29, 13.31, 14.1–14.8. Dimension the Link Arm using the complete decimal dimensioning system with metric values. If assigned, use decimal-inch equivalents. Include all necessary finish marks. Convert the following American Standard fits to metric values as required. The hole at **A** is to be reamed to a class FN 1 fit (see Appendix 9). Hole **B** is to be bored to a class RC 4 fit (see Appendix 5). Hole **C** is to be reamed to a class RC 1 fit (see Appendix 5). Hole **D** is to be reamed to receive a No. 2 American National Standard taper pin (see Appendix 24). Space all dimension lines 10 mm from the views and 10 mm apart.

Instructions

Drawing 66. Mating Parts Dimensioning. References: §§13.9, 13.20–13.27, 13.31, 14.1–14.8. Dimension the Base fully, including finish marks and notes. Space dimension lines 10 mm from the views and 10 mm apart. Use the complete decimal dimensioning system with metric values except for certain standard parts. If assigned, use decimal-inch equivalents. The large hole is to be reamed to metric values equivalent to an RC 6 fit (see Appendix 5). Note that the RC 6 fit also applies to the diameter of the Special Bolt on Drawing 67. The two small holes are to be drilled with 0.5 mm allowance and spotfaced for M8 × 1.25 Hexagon head cap screws.

Drawing 67. Mating Parts Dimensioning. References: §§13.9, 13.20–13.27, 13.31, 14.1–14.8, 15.21. These three parts fit with the Base on Drawing 66 and are to be dimensioned fully in the manner used on Drawing 66. Dimension the length of the Bushing with an allowance of 0.08 mm and a tolerance of 0.05 mm figured from the maximum length as the basic size. The tolerances should permit the Bushing to turn freely on the Special Bolt.

If assigned, draw an assembly of the Roller Guide with part identification numbers and a parts list. See Figure below for Hydraulic Grease Fitting details. References: §§16.14, 16.19–16.21.

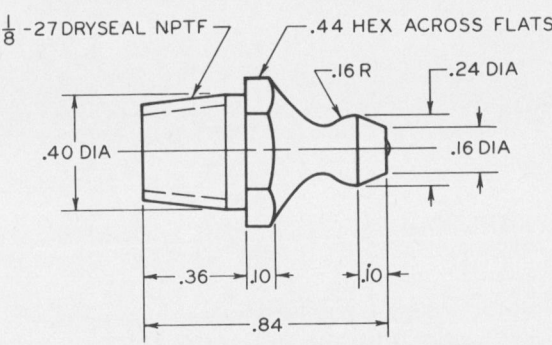

NO. 8585 HYDRAULIC GREASE FITTING
STEEL - BRIGHT ZINC PLATE

Drawing 68. Threads and Fasteners. References: §§15.1–15.21, 15.29–15.35. Draw light guide lines from the marks indicated, and letter the answers in the spaces provided. Standard abbreviations may be used to avoid crowding. See Appendix 4.

Drawing 69. Threads and Fasteners. References: §§15.7–15.10, 15.23–15.28; Appendix 22. Draw specified threads and fastener details, using the schematic thread symbols unless otherwise assigned. Complete the section lining and leaders where required. Chamfer ends of threads 45° × thread depth in Probs. 1, 2, and 3.

Drawing 70. Unified Threads. References: §§15.3–15.10, 15.20, 15.21.
Piston Rod. Complete the partial front view and the end view by drawing the threads specified.
Gland. Add the specified threads to the given views. Note that the front view is a half section, and the incomplete side view is a half view. Omit hidden lines in the sectioned front view, and add only the lines necessary to show the thread in the half side view. Complete the section lining in the sectional view.
Adjustable Link. This is an assembly of three parts, as indicated. The Link Base is threaded through. The Eye Bolt is engaged to the depth indicated, and its position locked by the Jam Nut. Complete the assembly view and also the half left-side view of the Link Base. Omit the other parts of the assembly in the half view. Complete all section lining in the sectional view.

In all three problems, complete the thread-note leaders.

Drawing 71. Detailed Acme and Square Threads. References: §§15.3–15.7, 15.12, 15.13, 15.15, 15.21.
Adjusting Screw. Draw the specified Acme threads to complete the view. Complete the leaders and add arrowheads touching the threads. Construct the threads so as to be symmetrical about the central neck of the screw.
Leveling Jack. Draw the specified square threads to complete the assembly view. Note that the scale of the drawing is double size. Add necessary section lining and complete the thread-note leader.

Drawing 72. Fasteners. References: §§15.7–15.10, 15.15, 15.23, 15.24, 15.29–15.31, Appendix tables 13, 14, 18, 20, and 22. Complete the sectioned assembly as indicated by drawing the specified American National Standard fasteners. Complete all required section lining. This assembly shows a Flexidyne dry fluid drive. Instead of fluid, a quantity of heat-treated shot is contained in the housing, which is keyed to the motor shaft. When the motor is started, centrifugal force throws the shot to the perimeter of the housing, packing it between the housing and the rotor that transmits power to the load.

Drawing 73. Welding Representation. References: §§27.1–27.10. In the upper portion of the sheet are shown six examples of welds accompanied by joints to which you are to attach the appropriate welding symbols with size specifications. In the lower portion of the sheet are three views to which you are to attach welding symbols for the welds indicated on the small isometric drawing.

Drawing 74. Pie Chart and Rectangular Graph.
Space 1. References: §§28.20, 28.21. According to the 1970 U.S. census, Chicagoans age 25 and over (1,943,464) reported the following educational achievement.

Never attended school	2%
Dropped out before grade 8	18%
Dropped out before grade 9	16%
Dropped out in high school	21%
Graduated from high school	26%
Dropped out of college	9%
Graduated from college	8%

Divide the pie chart to illustrate the above data. Show and label the appropriate percentages for each category. Use 3 mm (.12″) engineering lettering throughout. Balance the largest sector symmetrically about a

Instructions

vertical center line in the lower area of the circle. Title the chart: EDUCATIONAL EXPERIENCE OF CHICAGO RESIDENTS AGE 25 AND OVER—1970. Underline the title only. Indicate outside the area of the pie chart the total number that reported.

Space 2. References: §§28.15, 28.16. Construct a column or bar chart on the rectangular grid to show the auto accidents experienced by drivers of various age groups. The Motor Vehicle Mfrs. Assn. of U.S. Inc., reported the following data (1972).

Age Groups	Licensed Drivers	Drivers Involved in Fatal Accidents
Under 25	21.6%	33.8%
25–44	39.3%	37.7%
45–64	30.6%	21.1%
65 and over	8.5%	7.4%

Locate the axes at the given starting corner. Place the PERCENT scale on the Y-axis and the AGE GROUP scale along the X-axis. Two bars are required for each age group, similar to Fig. 28.31 (b). Cross-hatch at 45° the bars representing the licensed drivers. Indicate the percent values above all bars similar to Fig. 28.31 (a). Add border lines for the chart area and strengthen appropriate horizontal grid lines, but do not cross bar areas. Identify shading significance with key inserts similar to Fig. 28.31 (b). Title the chart: INVOLVEMENT IN FATAL ACCIDENTS, U.S. DRIVERS—1972.

Drawing 75. Arithmetic and Semilogarithmic Graphs.

Space 1. References: §§28.2–28.7. Complete the rectangular line chart as an engineering graph to represent the following data.

Safe-Load Capacities of Roller bearings Standard Medium Series

	Safe Load, lb	
RPM	Roller Size 9	Roller Size 12
100	1550	4200
500	1150	3100
1000	785	2100
1500	535	1450
2000	365	990
2500	250	680
3000	165	465

Draw a smooth curve through the plotted points. Add title and other necessary notation to the graph.

Space 2. References: §§28.8, 28.9. Construct a semilogarithmic graph to present the data given for problem in *Space 1*. Plot the dependent variable on the 2.00" cycle logarithmic scale. Draw smooth curves through the plotted points. Title the graph and add complete notation independently of problem in *Space 1*.

Observe the relationships of the data on the two graphs.

Drawing 76. Graphic Solutions of Simultaneous Equations.

Prob. 1. References: §§4.21–4.23, 31.1. Plot the two curves for the given simultaneous polynomial equations. From the intercept establish the graphic solutions to three significant figures. Show on your drawing a check of the solutions in the given equations. Label the solutions and the curves.

Prob. 2. Reference: §31.1. Plot the curves for the given simultaneous polynomial equations. Determine the values of the solutions to three significant figures. Check solutions in the equations and show proof on drawing. Label the solutions and the curves.

Drawing 77. Graphic Solutions of Roots of Equations.

Prob. 1. Reference: §31.1. Plot the given quadratic equation by the method of Fig. 31.1 (b). Determine the roots to three significant figures. Check roots in the equations and show on drawing. Label the roots and the curve.

Prob. 2. Reference: §31.1. A cubic equation will have either one or three real roots. Determine a sufficient number of points to insure an accurate plotting of the given expression. The intercepts on the X-axis provide the roots. Determine the root values to three significant figures. Check roots in equations and show on drawing. Label the roots and the curve.

Drawing 78. Graphic Solutions of Empirical Equations.

Prob. 1. References: §§30.1–30.4. Plot the straight line relationship of the mean temperature and the coefficient of expansion for Alloy-G on the given grid.

Temperature, °F	Coefficient of Expansion per °F $\times 10^{-6}$
160	5.68
530	6.08

Determine the empirical equation for the above relationship by the slope-intercept method. Show origin of data for the equation and the derived equation (correct to three significant figures) on the drawing. Check the solution by the selected-points method if assigned.

Prob. 2. Reference: 30.7. Plot the general relationship of tensile and compressive strengths for Alloy-G on the given logarithmic grid.

Pounds per Square Inch (psi)

Tensile Strength	Compressive Strength
100	500
135	600
200	820
300	1150
470	1500
720	2000
950	2500

Determine the empirical equation for the above relationship by the slope-intercept method, and check solution by selected-points method. Show proof on back side of sheet. Show origin of data for the equation and the derived equation (correct to three significant figures) on the drawing. Add appropriate title for the graph.

Drawing 79. Natural Parallel-Scale Nomograms.

Space 1. Reference: §29.8. Prepare a natural parallel-scale nomograph or alignment diagram for the given equation. Space the scales 3.00" apart. Make all scales *same length* as given X-scale. Match height of scale numerals to given numerals and identify each scale. Check the diagram by substituting the following values: $x = 0$, $y = 8$; $x = 12$, $y = 12$ and the appropriate Z-scale values in the equation. Show calculations on the drawing. Start Z-scale at 0. Maximum Y value is 22.

Space 2. References: §§29.14, 29.15. Prepare a natural scale N-chart for the Ohm's law equation, $V = IR$, where V = voltage, I = current (scale range 0 to 10 amperes), and R = resistance (scale range 0 to 100 ohms). Make the vertical scales of equal length and spaced 6.00" apart. Match scale numeral heights to given numerals and label the scales for volts and ohms. Check the diagram by substituting the following values: 8 amperes, 50 ohms; 2.5 amperes, 80 ohms, and the appropriate voltage values in the equation. Show these calculations on the drawing.

Drawing 80. Graphical Calculus—Differentiation. References: §§31.2–31.6. An experimental rocket engine consumes its liquid fuel supply according to the given data. Plot these data as a smooth curve on the given X- and Y-axes grid.

Time, sec	Fuel, liters	Time, sec	Fuel, liters
0	0	35	810
5	45	40	865
10	150	45	900
15	315	50	920
20	510	55	935
25	650	60	940
30	745		

Determine the first derivative of the initial curve, and plot these data on the X'- and Y'-axes grid. The first derivative will give the rate of fuel consumption at any instant. Use the slope method of Fig. 31.4 unless otherwise assigned. Establish the base of each slope triangle as a value of 10 on the X-axis. The tangent of the slope angle is then one-tenth the value of the ordinate on the Y-axis.

Therefore, if the scale on the Y'-axis is expanded in a 10:1 ratio to the Y-axis, the ordinate of the slope triangle may be transferred directly to the construction for the first derivative.

Indicate the rate of fuel consumption 30 seconds after firing in the space at the bottom of the sheet.

Drawing 81. Graphical Calculus—Integration. References: §§31.2, 31.12, 31.13. The given curve of volume vs pressure is characteristic of data often encountered in science and engineering. The integral in this case is the work curve (in.-lb vs in.^{3}) and is equivalent to the area under the given curve. The work unit is inch-pound.

Determine the integral curve. Use the area-law method of Fig. 31.12 unless otherwise assigned. Establish the necessary axes and grids with appropriate labeling in the space provided. Cross-hatch the areas used to establish the mean ordinates. Show lengths of mean ordinates and summations of ordinates used to establish the integral curve. Also show how one point on the integral curve was located.

Drawing 82. Computer-Aided Drafting: Terms and Descriptions. References: Chapters 3 and 8, Appendix 3. Some terms related to computer graphics are given in the table. A list of descriptions for these terms is given on the right. Find the matching description for each term and enter its letter identifier in the table.

Drawing 83. Computer-Aided Drafting: Two-Dimensional Coordinate Plot. Reference: Chapter 8.

Space 1. Digitize the single view drawing by defining the X and Y coordinates of the indicated points and fill in the given table. Point A is the origin with values of X and Y equal to zero. Consider each division of the grid as 1 unit. Keep in mind that any X values to the left of the origin and any Y values below the origin are negative.

Space 2. From the X and Y coordinate date given in the table, plot all points on the grid and complete the drawing. Point A is the origin. Consider each division of the grid as 1 unit.

Drawing 84. Computer-Aided Drafting: Three-Dimensional Coordinate Plot. Reference: Chapter 8. In drawing an image, the actions of the pen are Move and Draw.

Move: The pen moves from its present position to new X, Y, and Z coordinates specified. A line is not drawn. Numeral 0 is used to indicate Move action.

Draw: A line is drawn from the present pen position to new X, Y, and Z coordinates specified. Numeral 1 is used to indicate Draw action.

Space 1. Determine X, Y, and Z coordinates for all the points of the object. Complete the table for drawing the object, starting with point A. Coordinates X, Y, and Z are positioned as indicated by the arrows, with point A as origin. Try to use a minimum number of Move actions.

Space 2. According to the data shown in the table, draw the object on the grids provided. Coordinates X, Y, and Z are positioned as indicated by the arrows, with point A as origin.

Drawing 85. Computer-Aided Drafting: Menu Usage. Reference: Chapter 8. The drawing shows the front view of a Bracket that is to be generated on a graphics terminal. The numbers 1 to 21 refer to graphic entities that make up the drawing. Available menu commands for generating enti-

Instructions

ties are given on the right. Complete the table by determining the menu commands to generate the entities. Enter the letter identifiers (A, B, etc.) of menu selections in the table.

Drawing 86. Computer-Aided Drafting: Coordinate Systems. Reference: Chapter 8. Using the given descriptions for VIEW COORDINATES and WORLD COORDINATES, complete the tables for the front and right-side views of the object. Point number 1 is considered as the origin. Each grid division is equal to 1 unit.

Drawing 87. Detail Drawings. References: Chapters 6–10, 13-15, §16.8. Draw or sketch the necessary views of the object assigned. Select appropriate scale and sheet size. Dimension completely using metric or decimal-inch dimensions as assigned.

Alternate Assignment: Using a CAD system, produce a hard-copy multiview drawing of the problem assigned. Dimension completely.

Drawing 88. Detail Drawings. References: Chapters 6–10, 13-15, §16.8. Draw or sketch the necessary views of the object assigned. Select appropriate scale and sheet size. Dimension completely using metric or decimal-inch dimensions as assigned.

Alternate Assignment: Using a CAD system, produce a hard-copy multiview drawing of the problem assigned. Dimension completely.

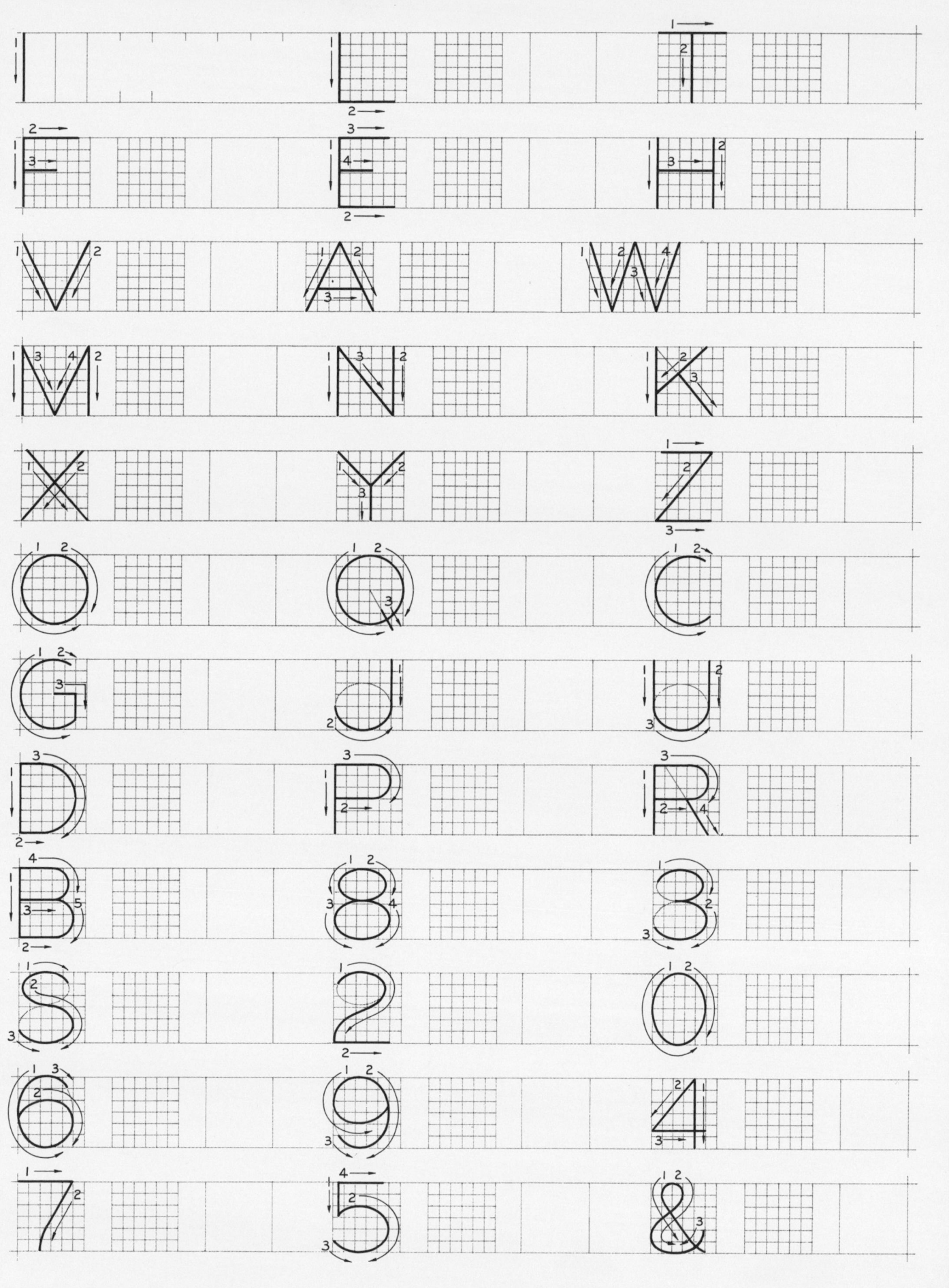

| CAPITALS AND NUMERALS VERTICAL LETTERING | DRAWN BY | FILE NO. | DRAWING 1 |

WHILE IT IS TRUE THAT

"PRACTICE MAKES PERFECT," IT

MUST BE UNDERSTOOD THAT

PRACTICE IS NOT ENOUGH, BUT IT

MUST BE ACCOMPANIED BY A CON-

TINUOUS EFFORT TO IMPROVE. EXCEL-

LENT LETTERERS ARE OFTEN NOT GOOD

WRITERS. USE A FAIRLY SOFT PENCIL, AND AL-

WAYS KEEP IT SHARP, ESPECIALLY FOR SMALL

LETTERS. MAKE THE LETTERS CLEAN-CUT AND

DARK-NEVER FUZZY, GRAY, OR INDEFINITE. 1234

$\frac{1}{2}$ 1.500 $\frac{3}{16}$ 45'-6 32° 15.489 $\frac{13}{64}$ 12"=1'-0 7$\frac{5}{16}$ 12.3 $\frac{1}{2}$ 2$\frac{1}{4}$

ONE MUST HAVE A CLEAR MENTAL IMAGE OF THE LETTERS. 234

| CAPITALS AND NUMERALS
VERTICAL LETTERING | DRAWN BY | FILE NO. | DRAWING
2 |

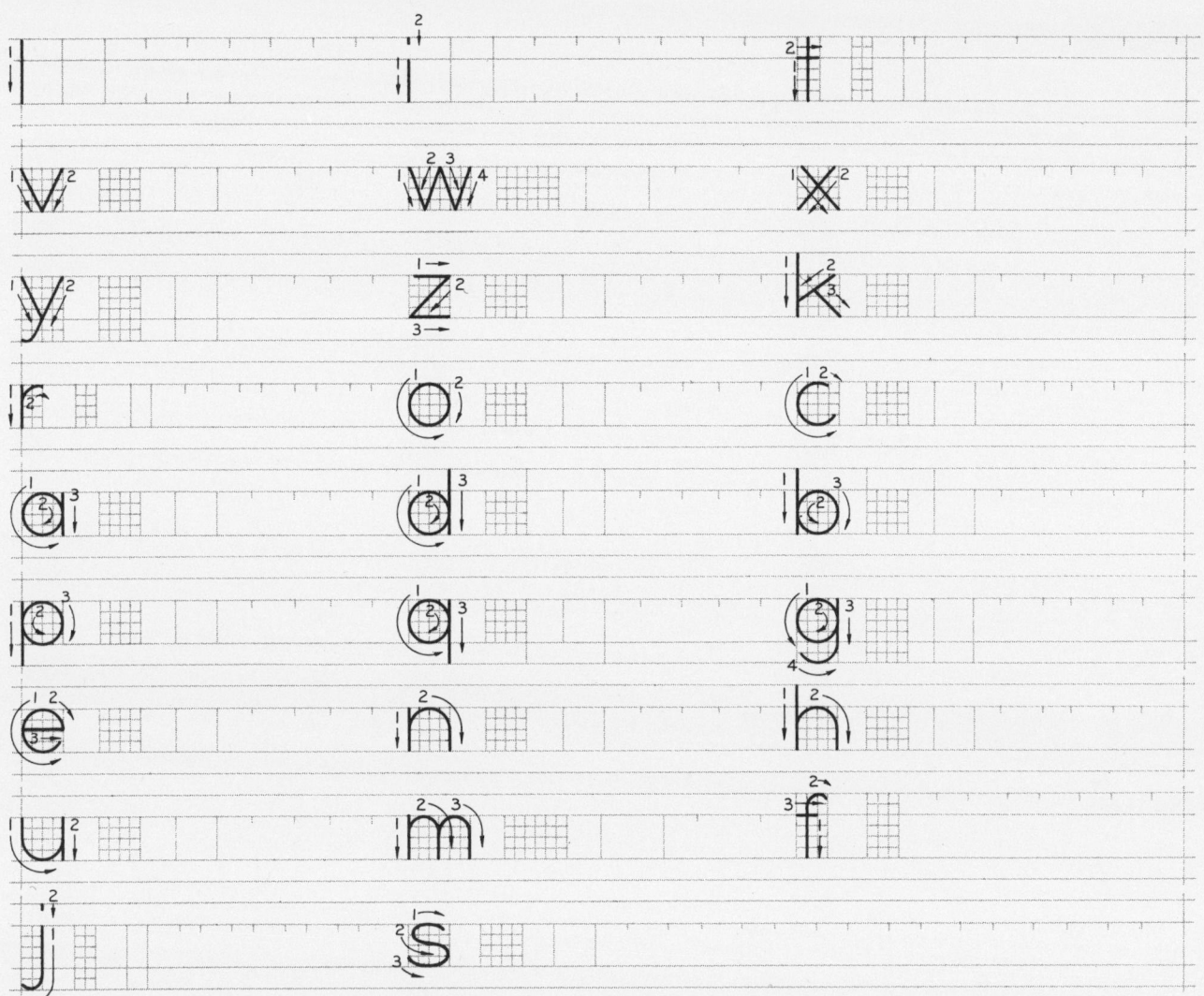

Now they are universally used for engineering

working drawings. They should not be executed

mechanically but should be made entirely freehand. The term "single-

stroke" does not imply that the letter is produced with a continuous

movement of the pencil or pen, but that the letter is made of one

LOWER CASE
VERTICAL LETTERING

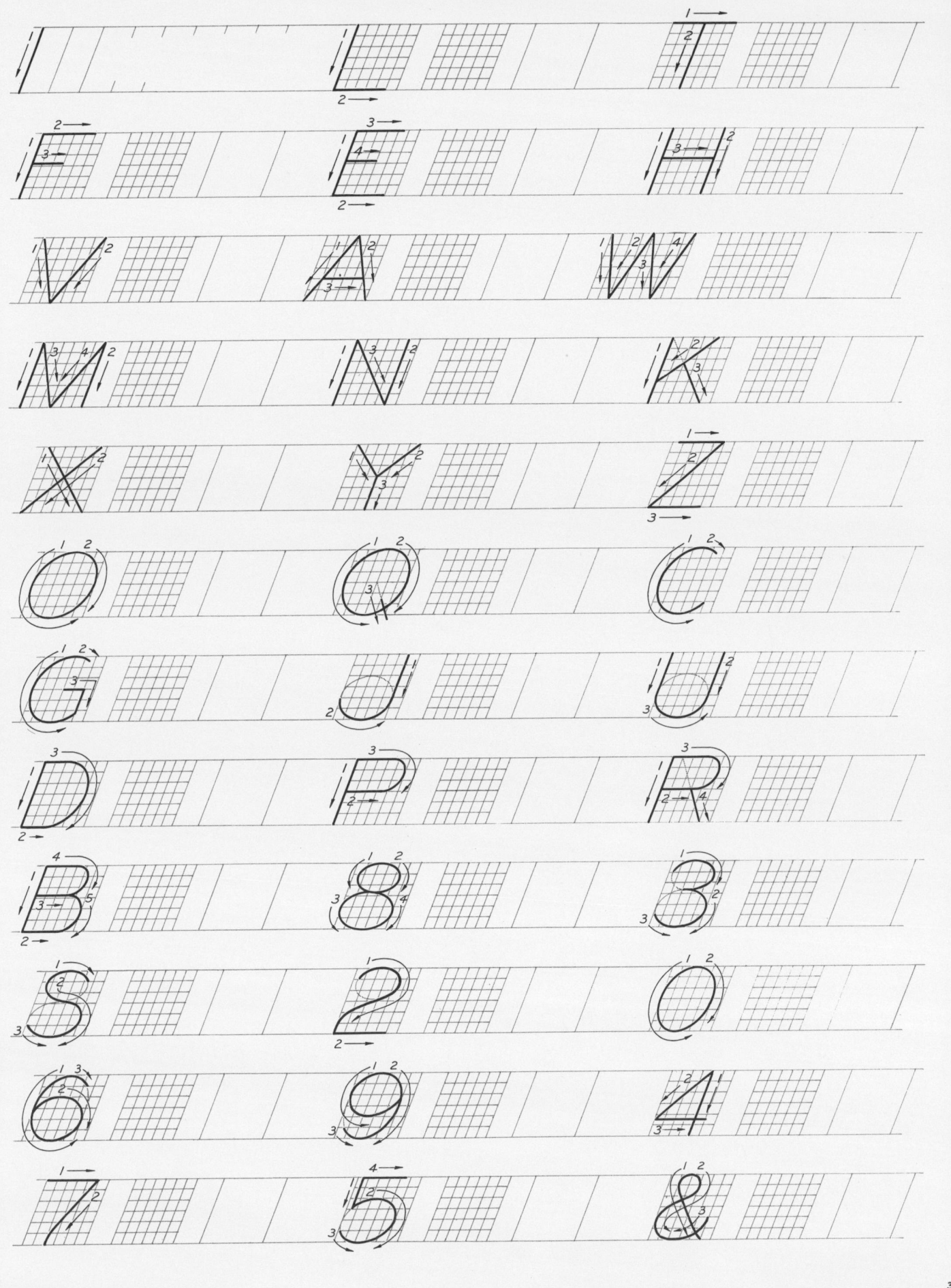

CAPITALS AND NUMERALS
INCLINED LETTERING

DRAWN BY

FILE NO.

DRAWING 5

WHILE IT IS TRUE THAT

"PRACTICE MAKES PERFECT," IT

MUST BE UNDERSTOOD THAT

PRACTICE IS NOT ENOUGH, BUT IT

MUST BE ACCOMPANIED BY A CON-

TINUOUS EFFORT TO IMPROVE. EXCEL-

LENT LETTERERS ARE OFTEN NOT GOOD

WRITERS. USE A FAIRLY SOFT PENCIL, AND AL-

WAYS KEEP IT SHARP, ESPECIALLY FOR SMALL

LETTERS. MAKE THE LETTERS CLEAN-CUT AND

DARK-NEVER FUZZY, GRAY, OR INDEFINITE. 1234

$1\frac{1}{2}$ 1.500 $\frac{3}{16}$ 45'-6 32° 15.489 $\frac{13}{64}$ 12"=1'-0 $7\frac{5}{16}$ 12.3 $\frac{1}{2}$ $2\frac{1}{4}$

ONE MUST HAVE A CLEAR MENTAL IMAGE OF THE LETTERS. 234

| CAPITALS AND NUMERALS INCLINED LETTERING | DRAWN BY | FILE NO. | DRAWING 6 |

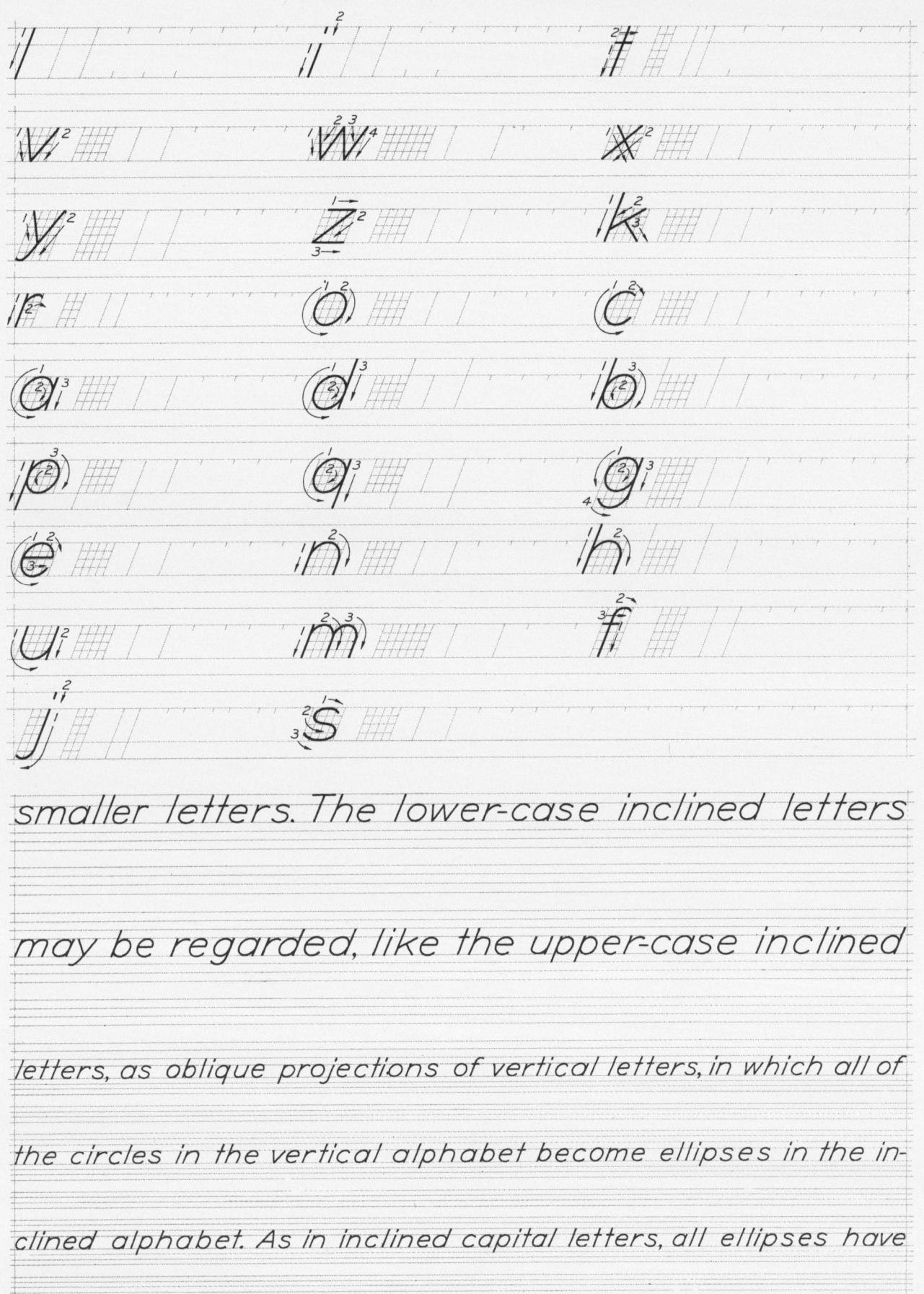

smaller letters. The lower-case inclined letters

may be regarded, like the upper-case inclined

letters, as oblique projections of vertical letters, in which all of

the circles in the vertical alphabet become ellipses in the in-

clined alphabet. As in inclined capital letters, all ellipses have

LOWER CASE INCLINED LETTERING

1	2

VISIBLE LINE

HIDDEN LINE

SECTION, DIMENSION, AND EXTENSION LINE

CENTER LINE

CUTTING-PLANE LINE

PHANTOM LINE

VISIBLE LINE
HIDDEN LINE
SECTION, DIMENSION, AND EXTENSION LINE
CENTER LINE
CUTTING-PLANE LINE
PHANTOM LINE
SHORT-BREAK LINE

3	4
5	6

ALPHABET OF LINES
MECHANICAL DRAWING

DRAWN BY | FILE NO. | DRAWING **9**

1

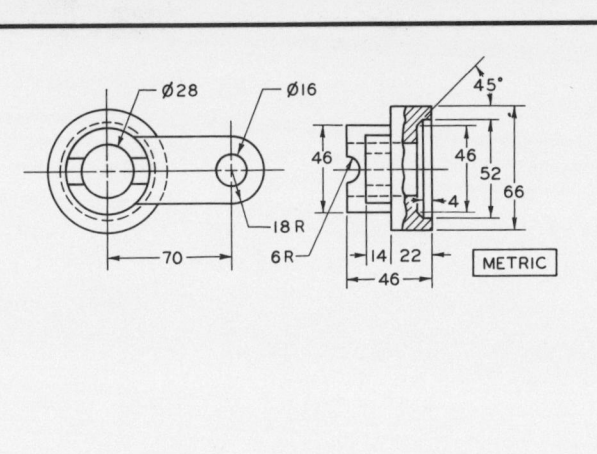

INDEX ARM

C 1—2 REQD
FULL SIZE

Draw the two views below.
Omit dimensions.

2

AGE STRUCTURE
U.S. RESIDENT POPULATION
1970

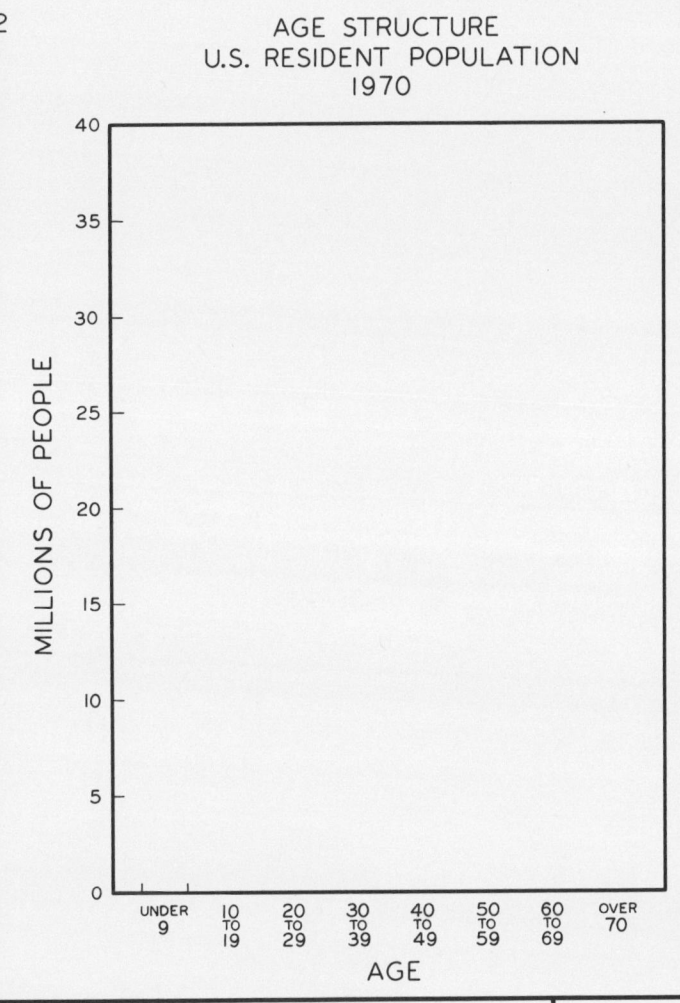

3

INFLUENCE OF CUTTING SPEED
ON MACHINING COST

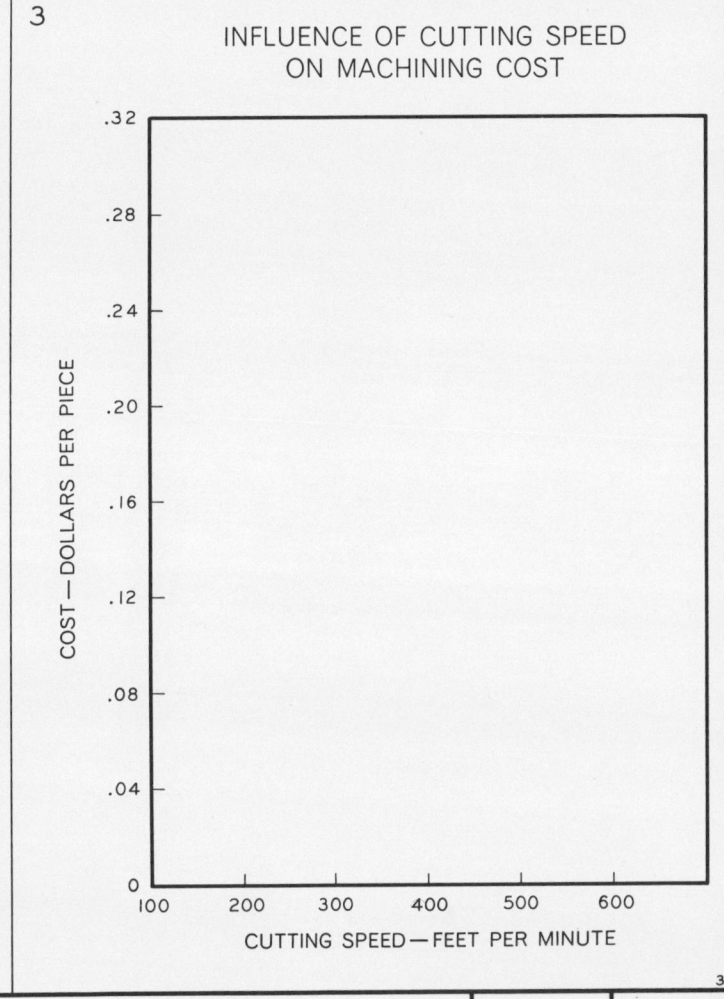

| LAYOUT, CHART, AND GRAPH | DRAWN BY | FILE NO. | DRAWING |
| MECHANICAL DRAWING | | | 11 |

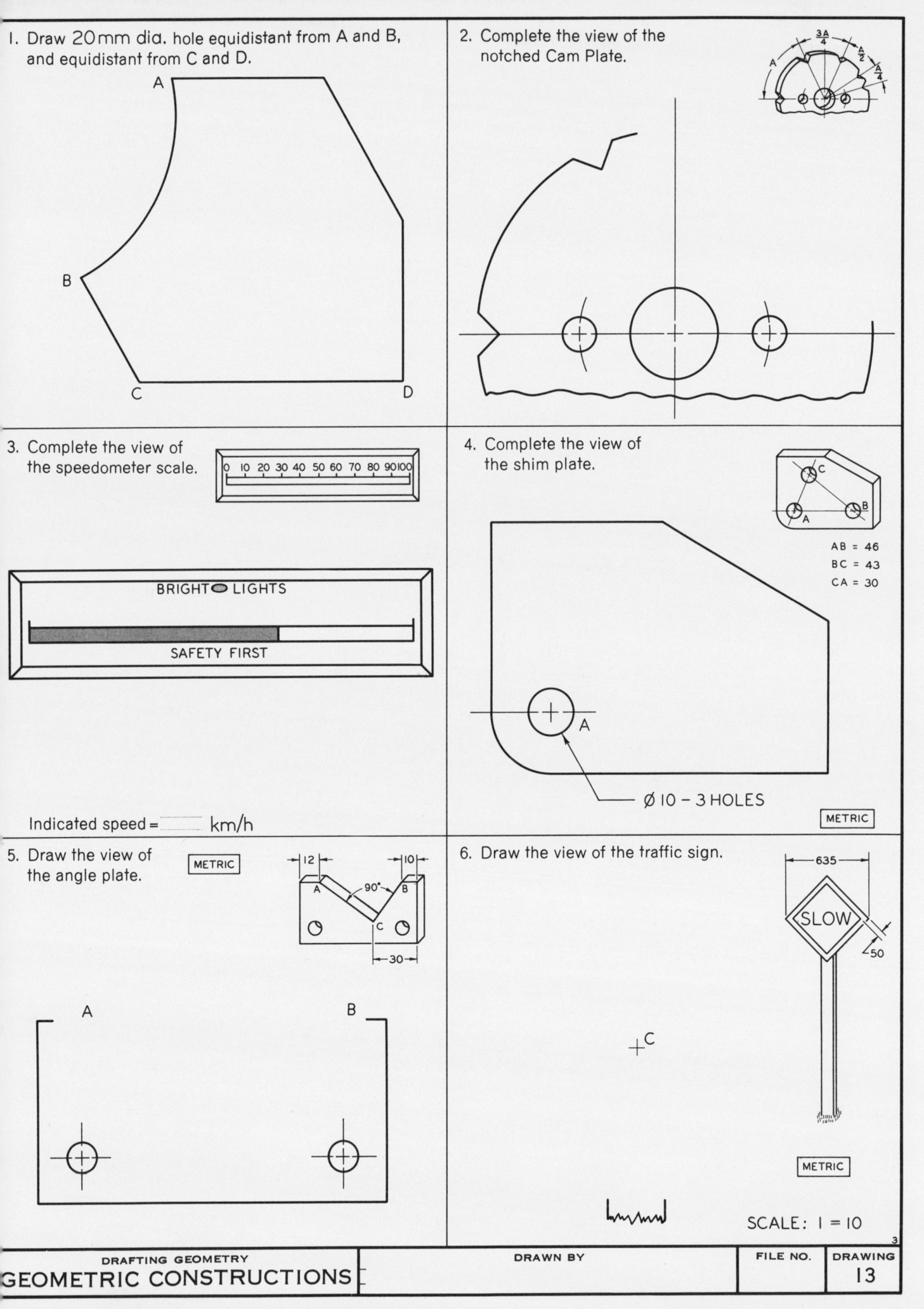

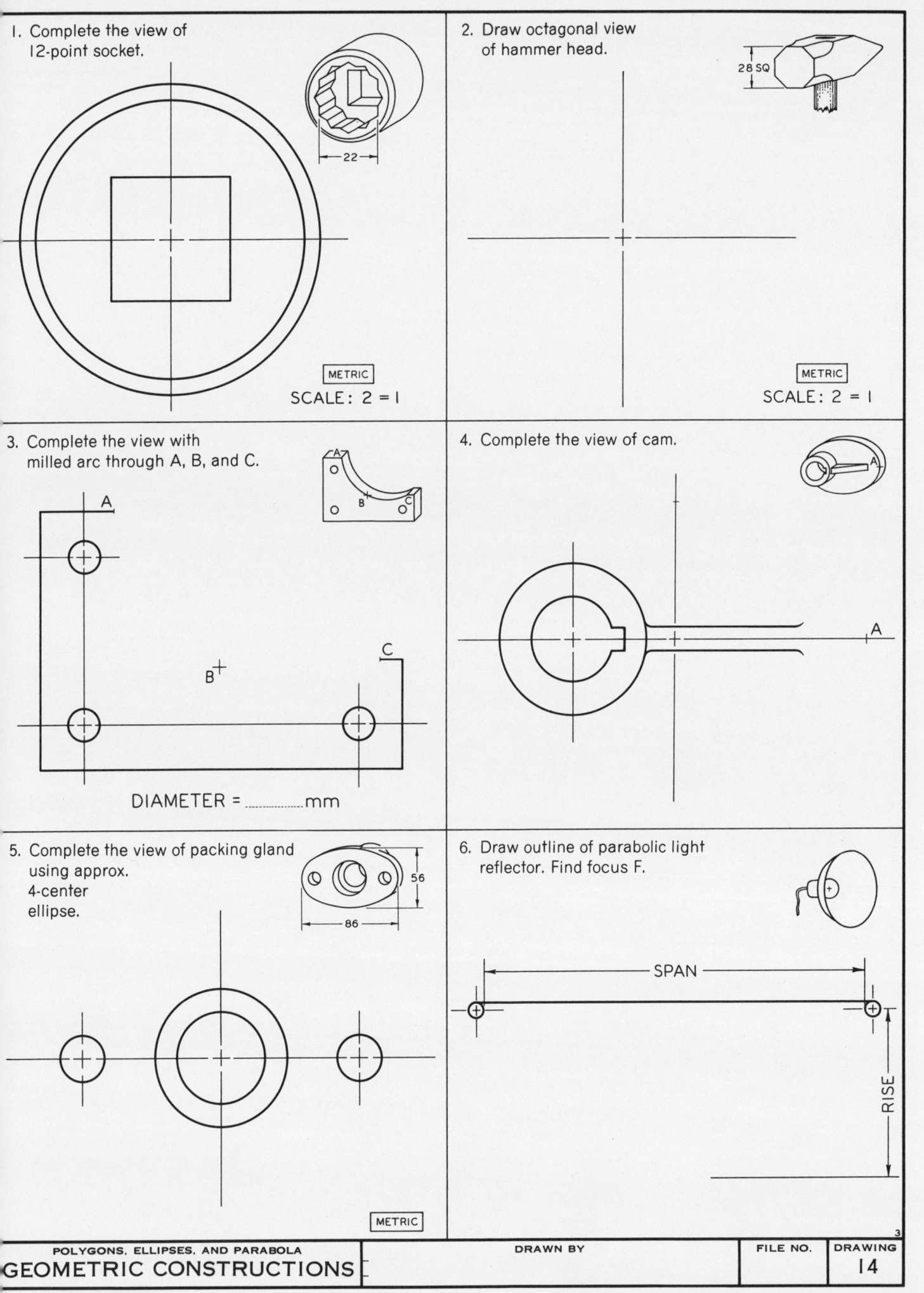

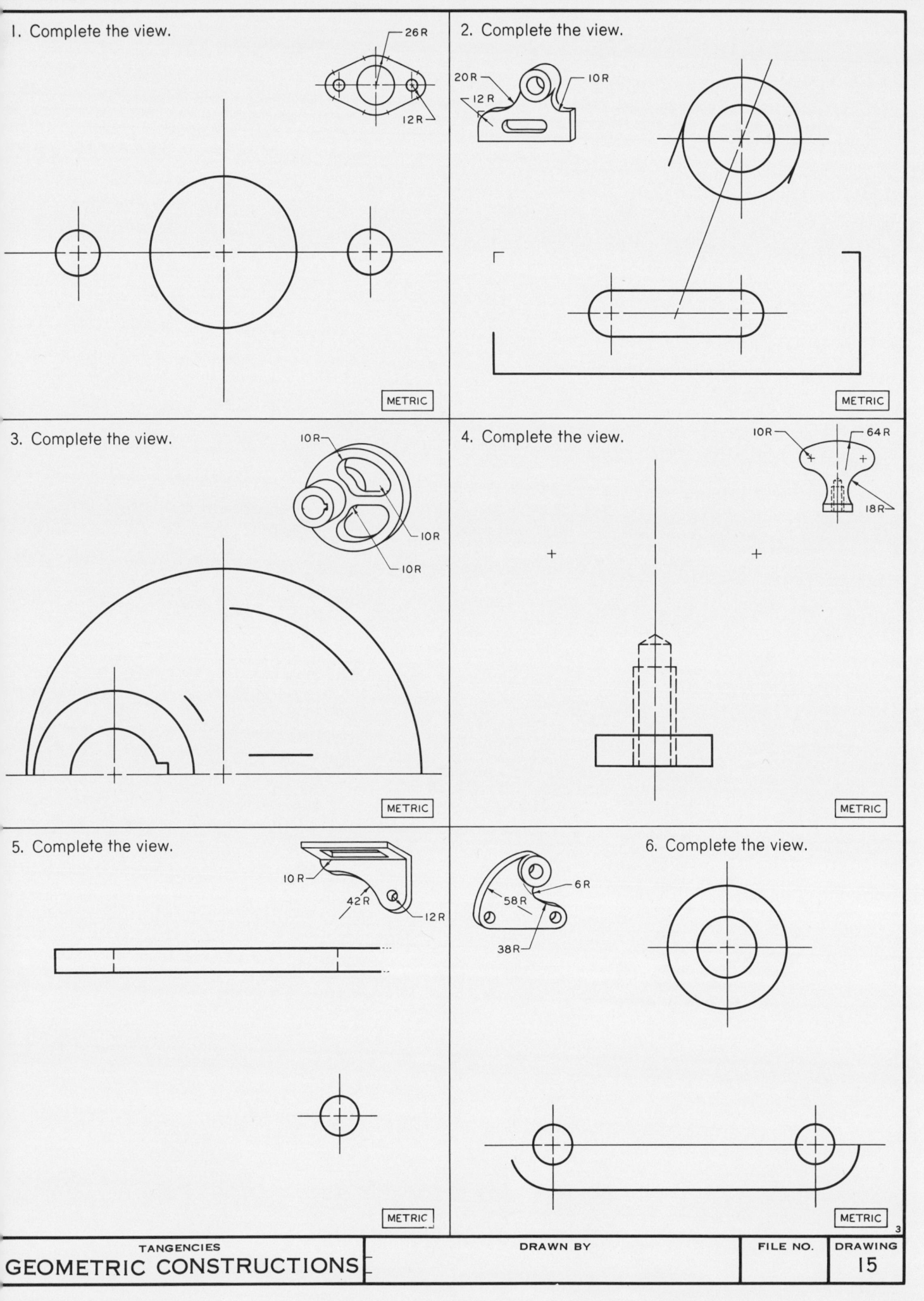

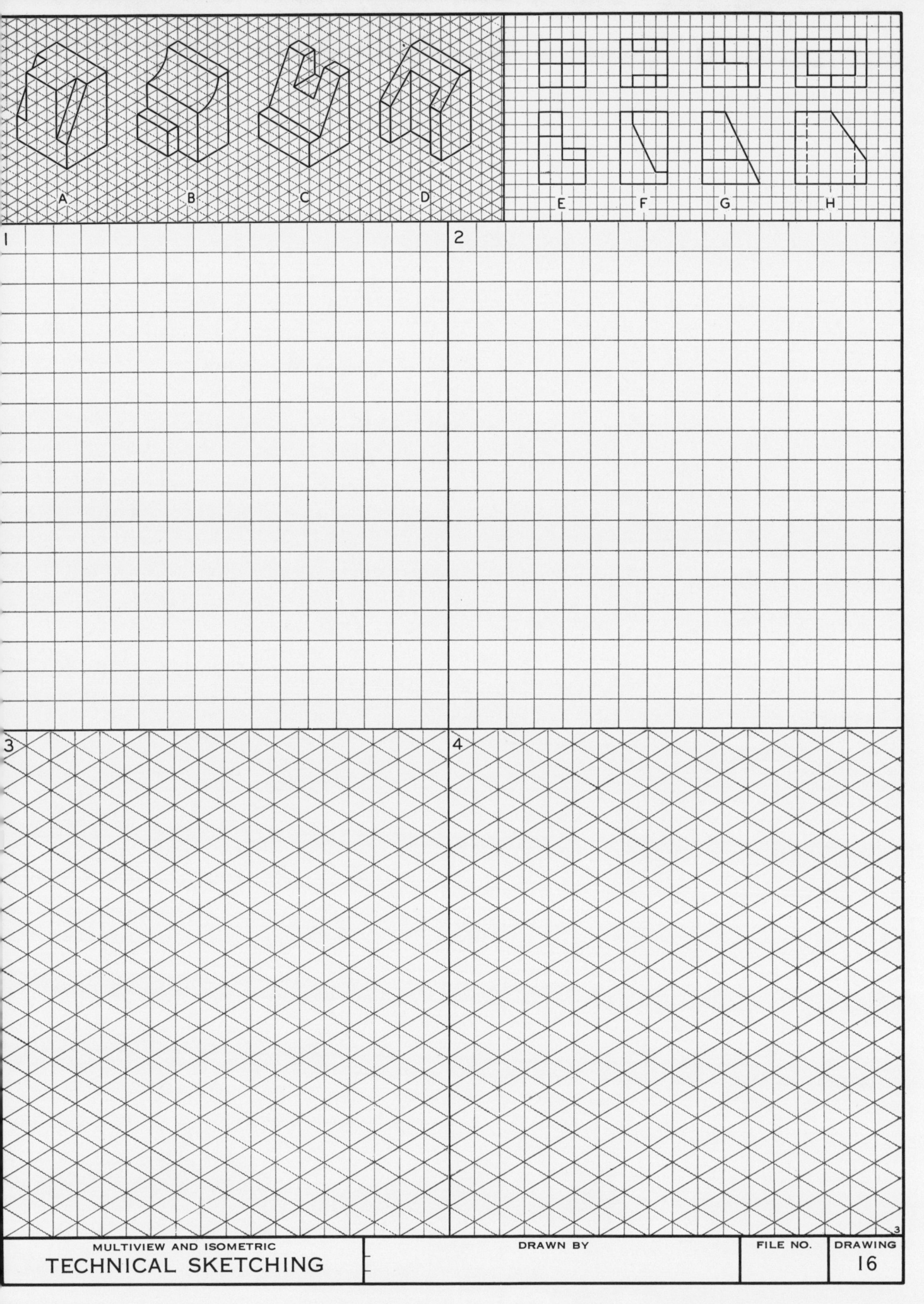

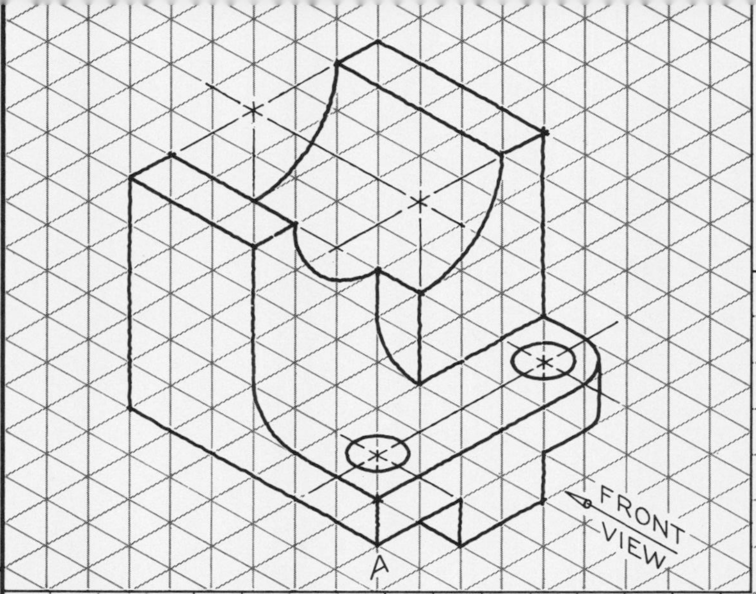

SPECIAL BEARING

Make a full-size 6-view freehand sketch of the object, with the views in the standard arrangement. Label all views: FRONT, TOP, etc. Space the views two squares apart. Show all hidden lines and center lines.

FRONT

FRONT VIEW

The necessary views for a complete shape description are:

MULTIVIEW
TECHNICAL SKETCHING

DRAWN BY

FILE NO.

DRAWING
17

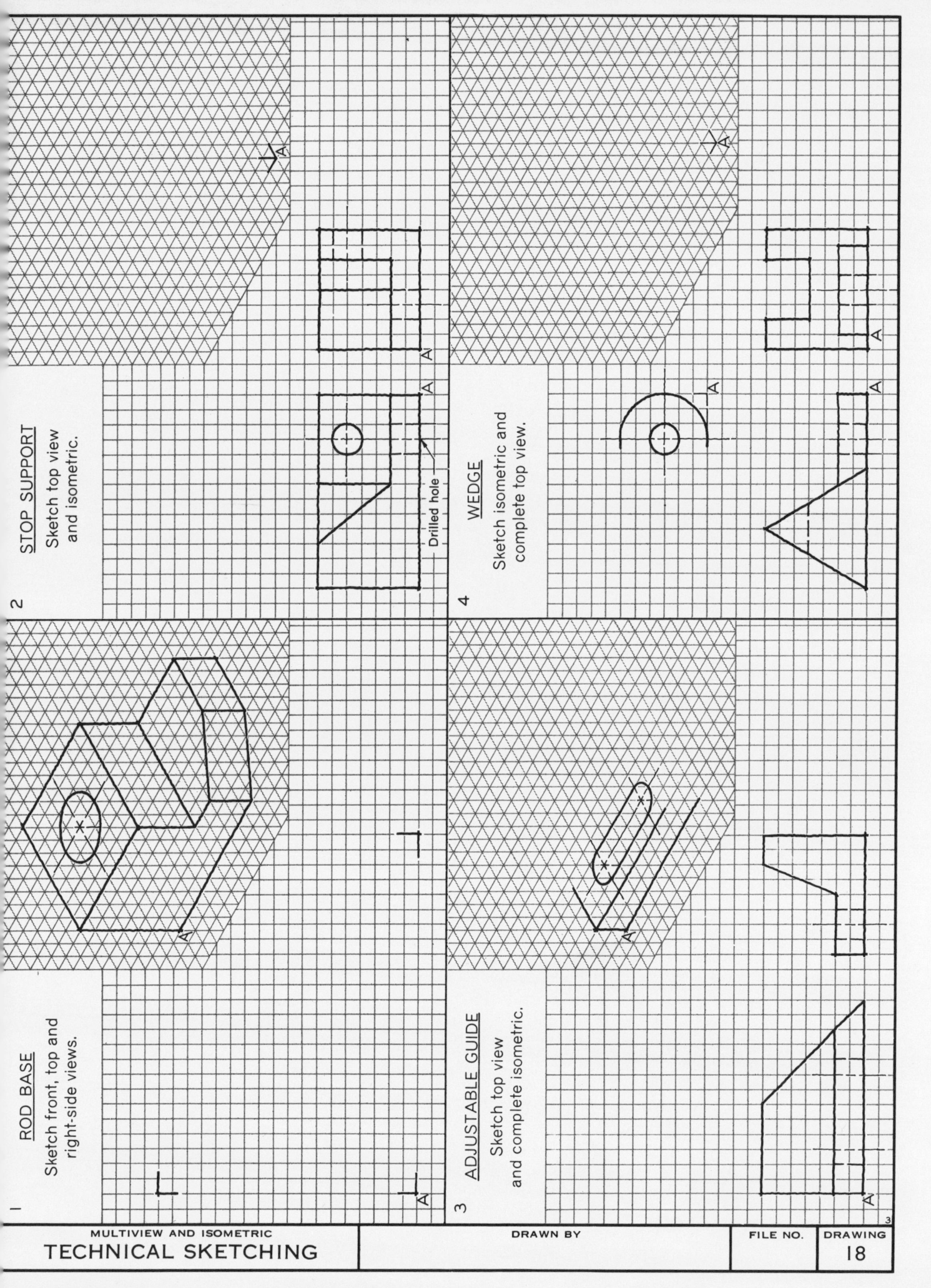

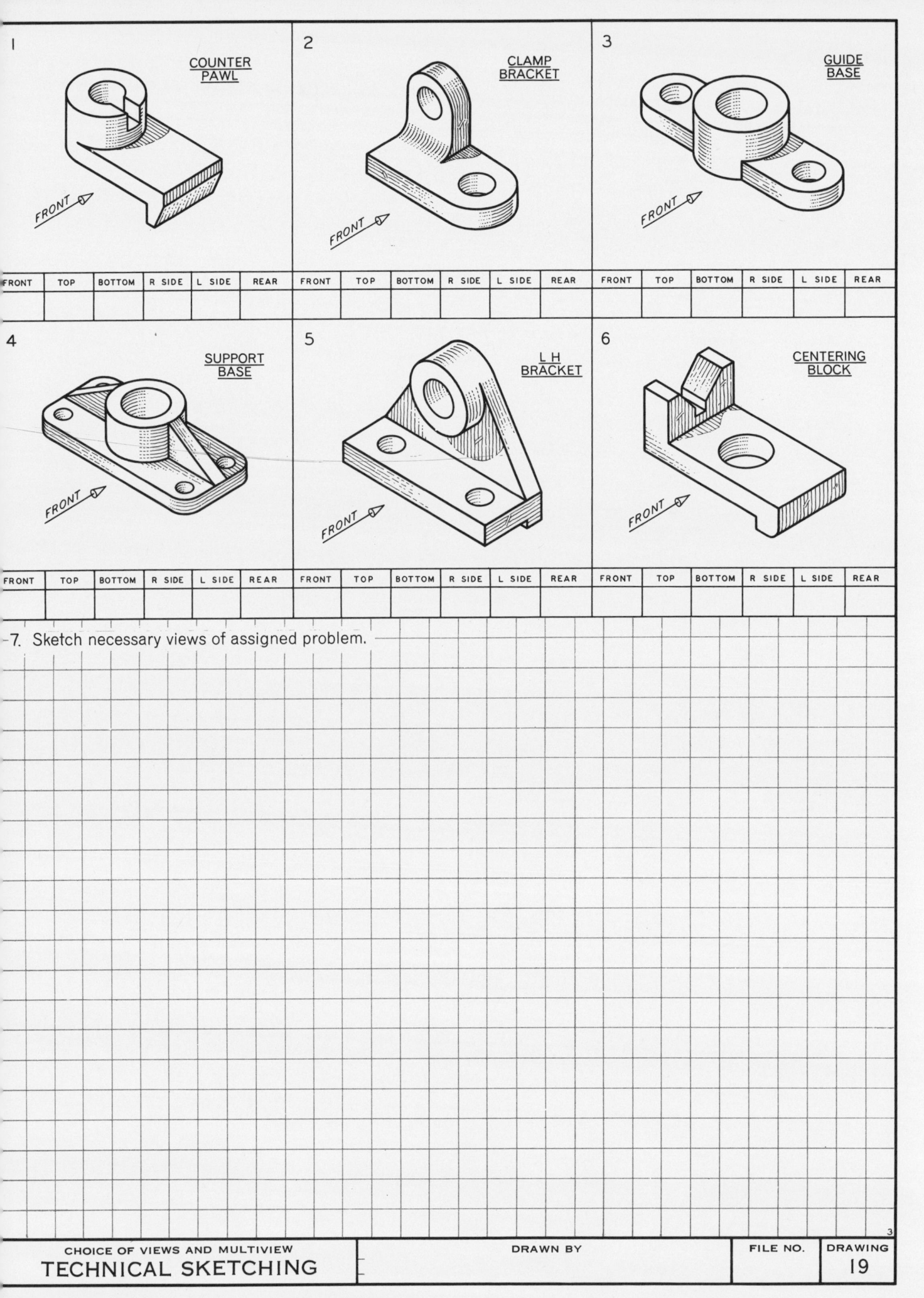

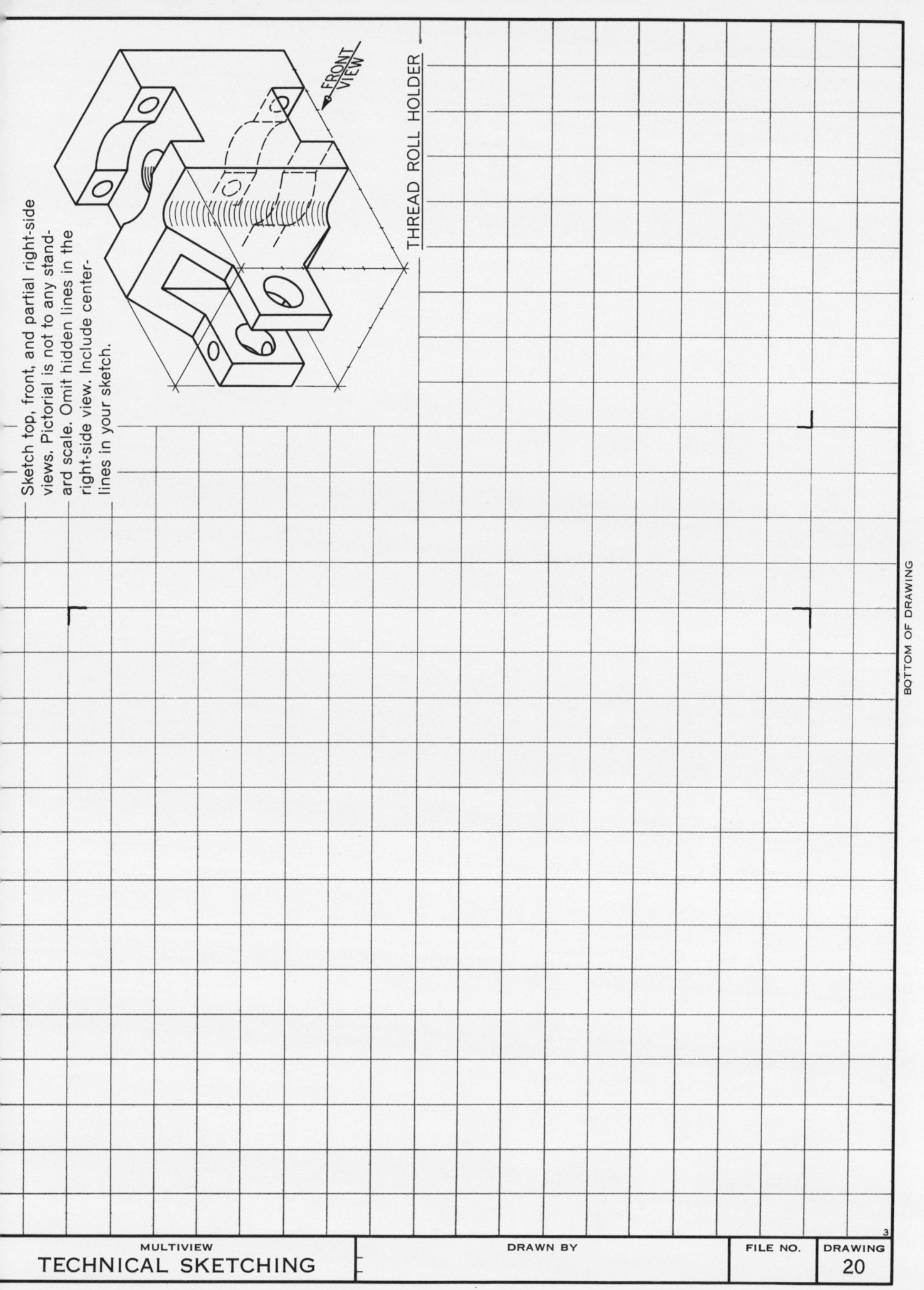

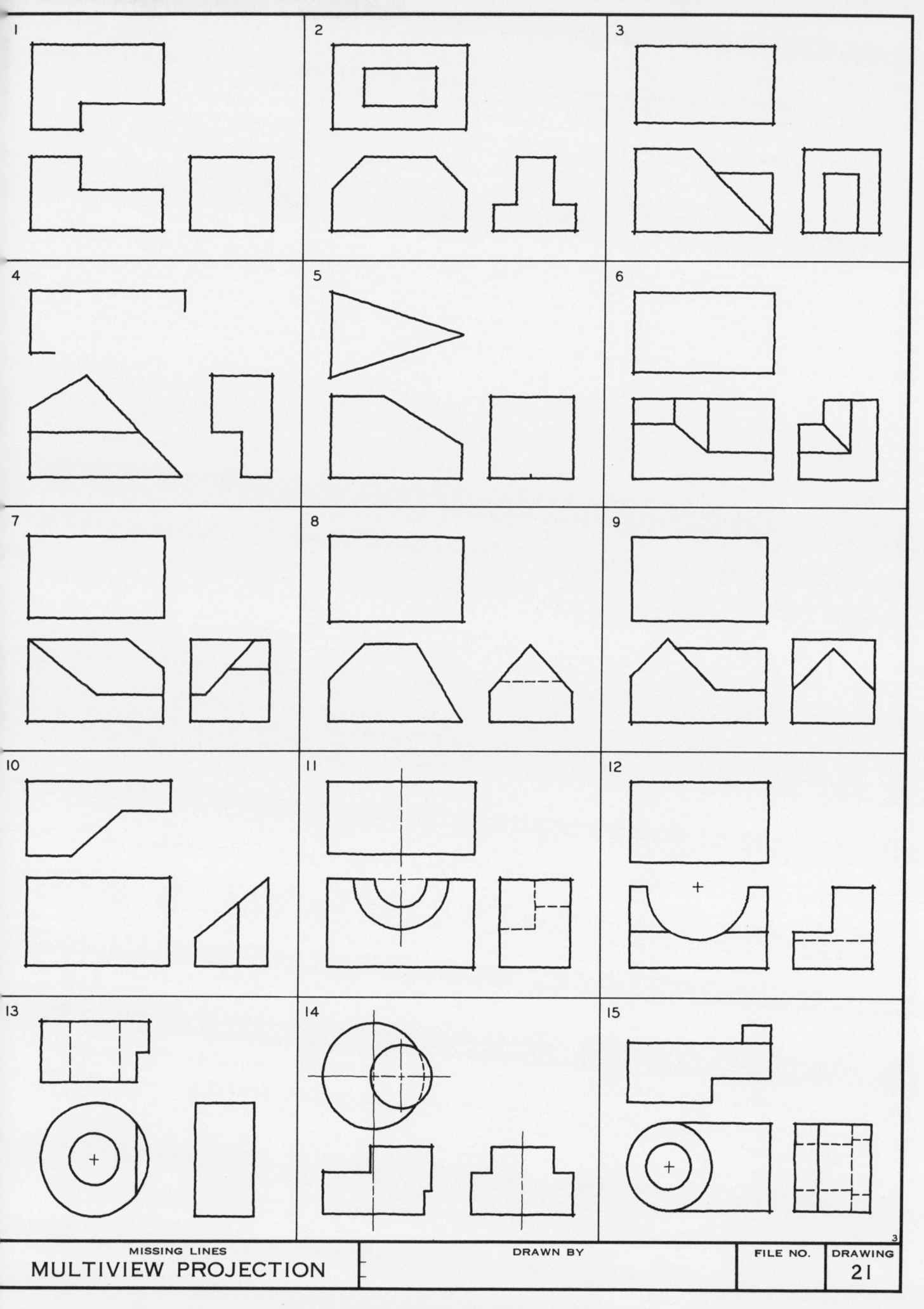

MISSING LINES
MULTIVIEW PROJECTION
DRAWING 21

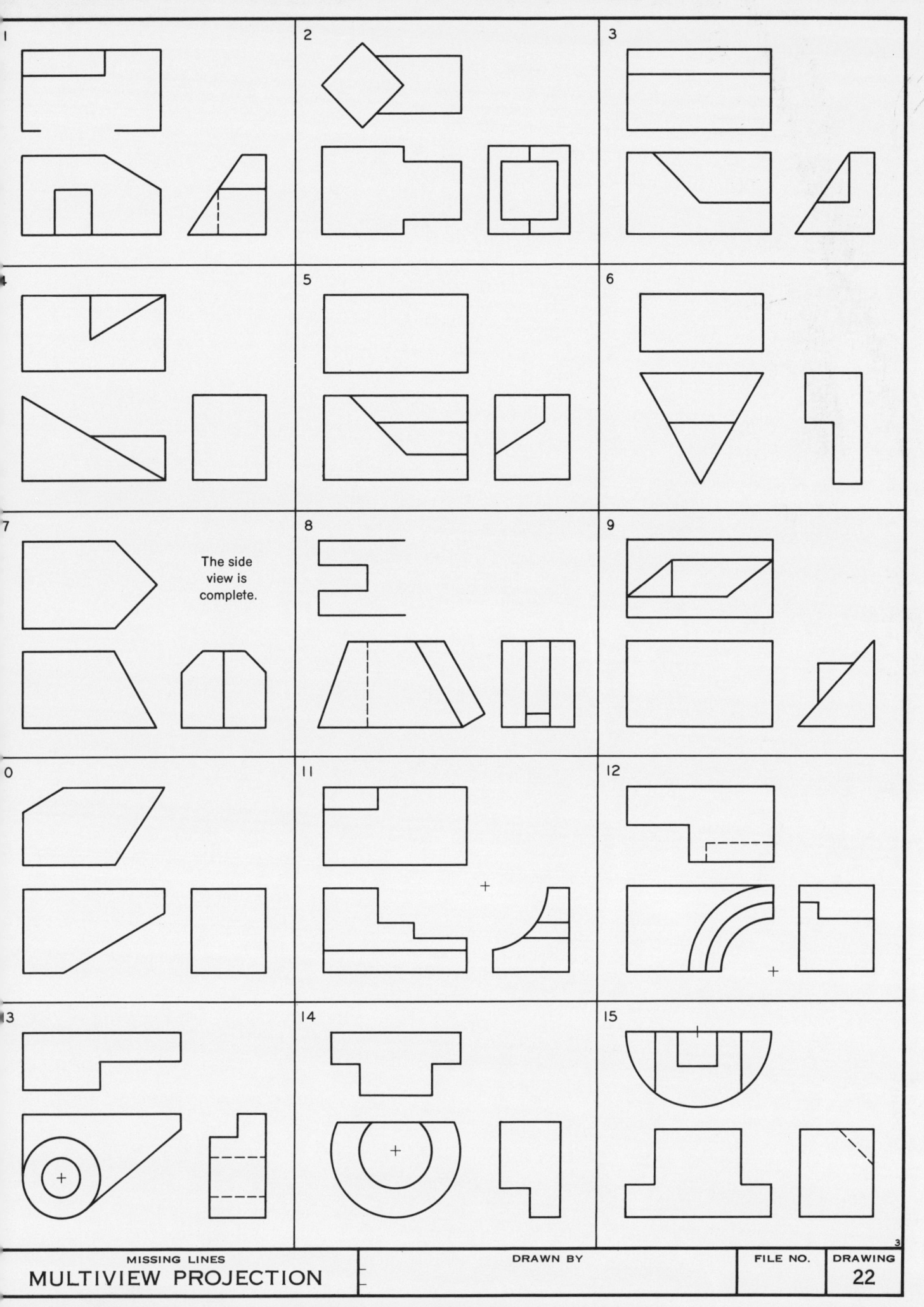

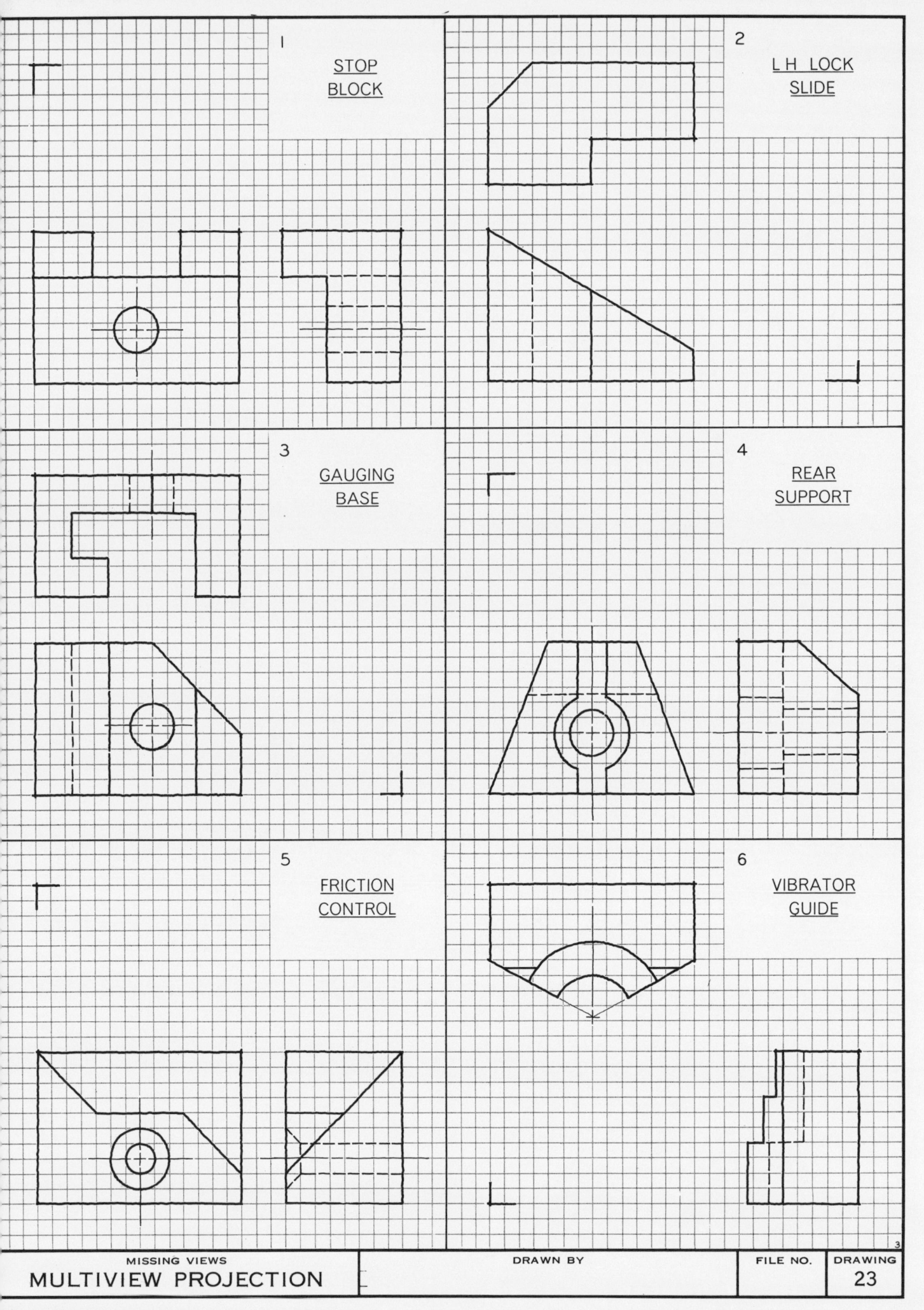

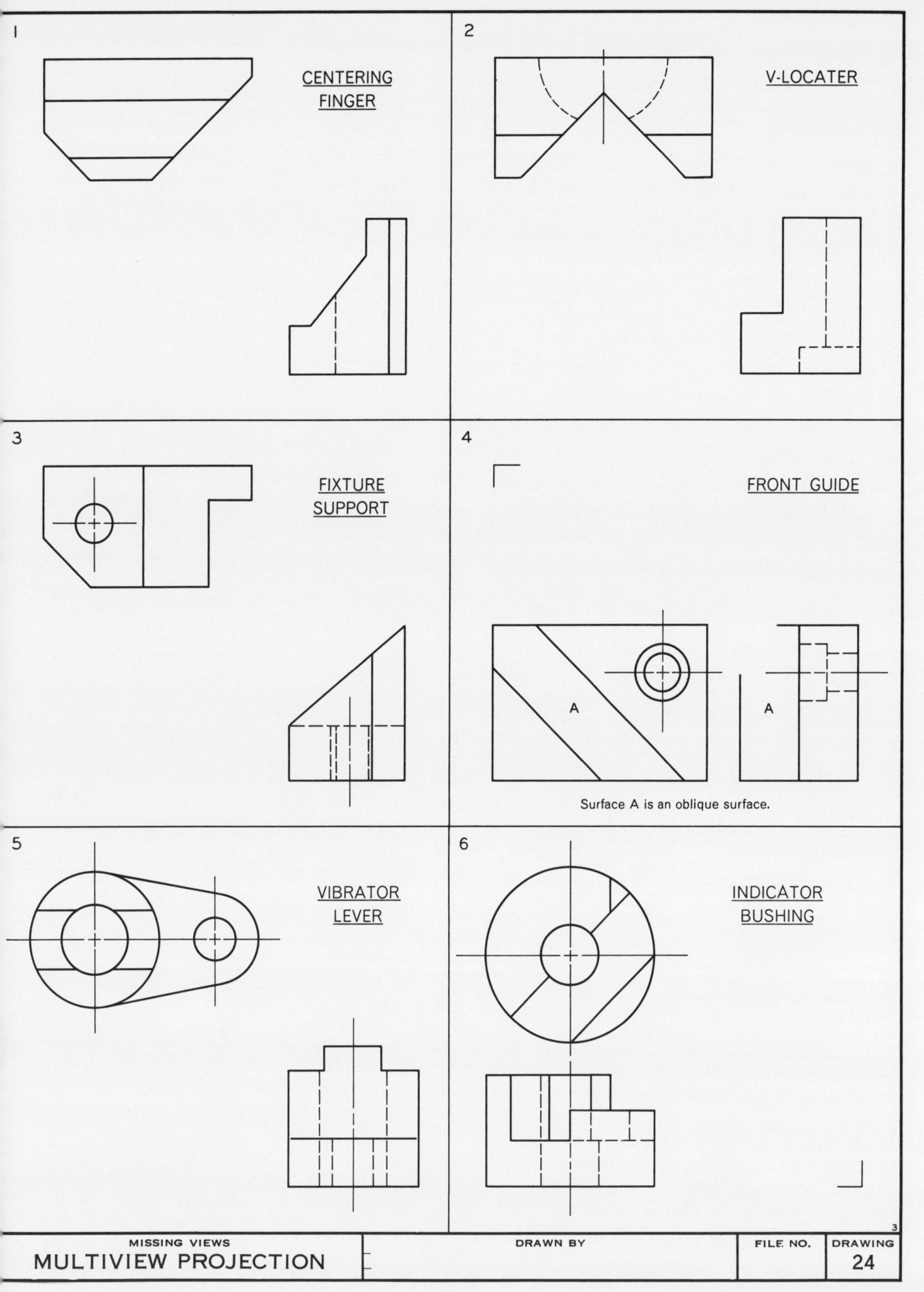

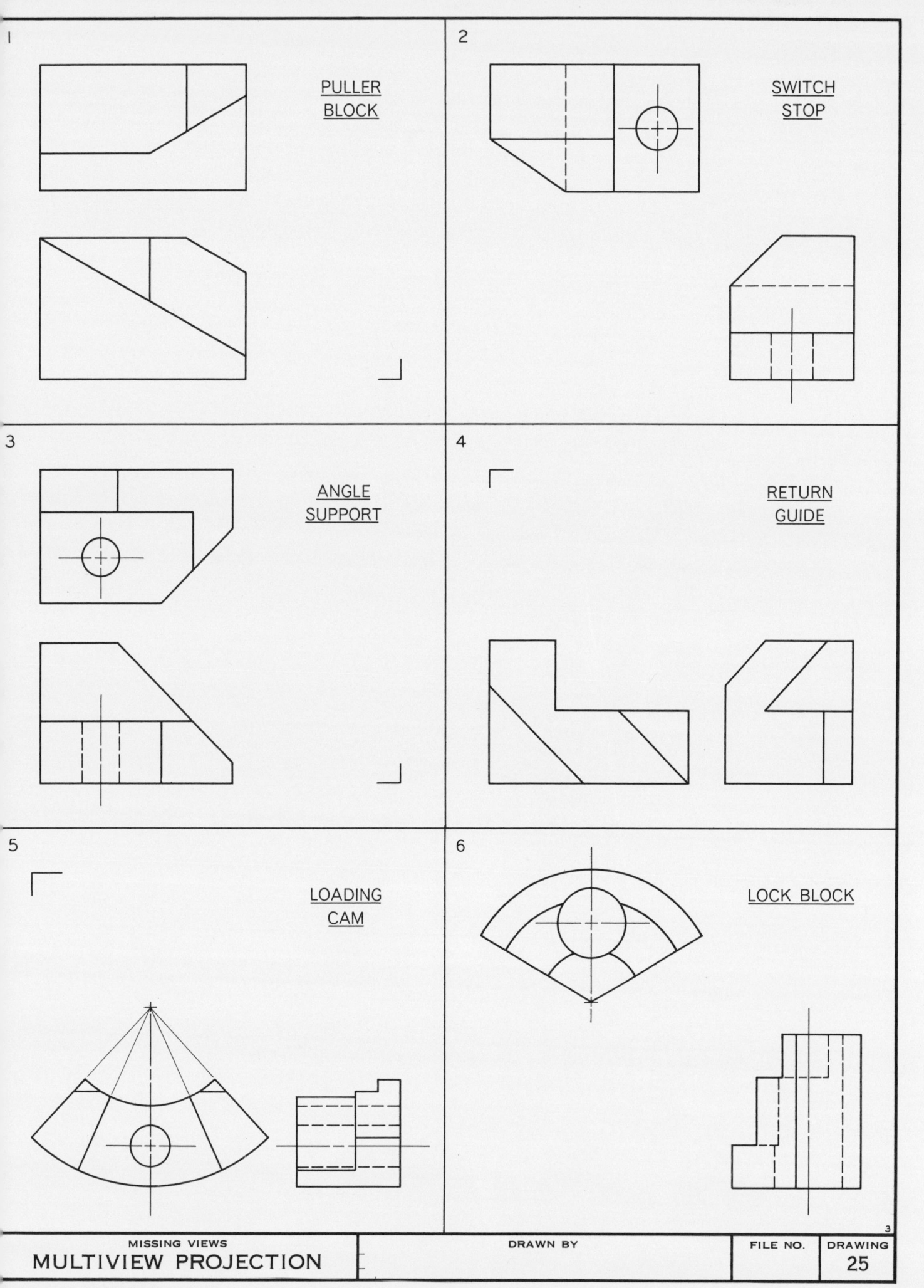

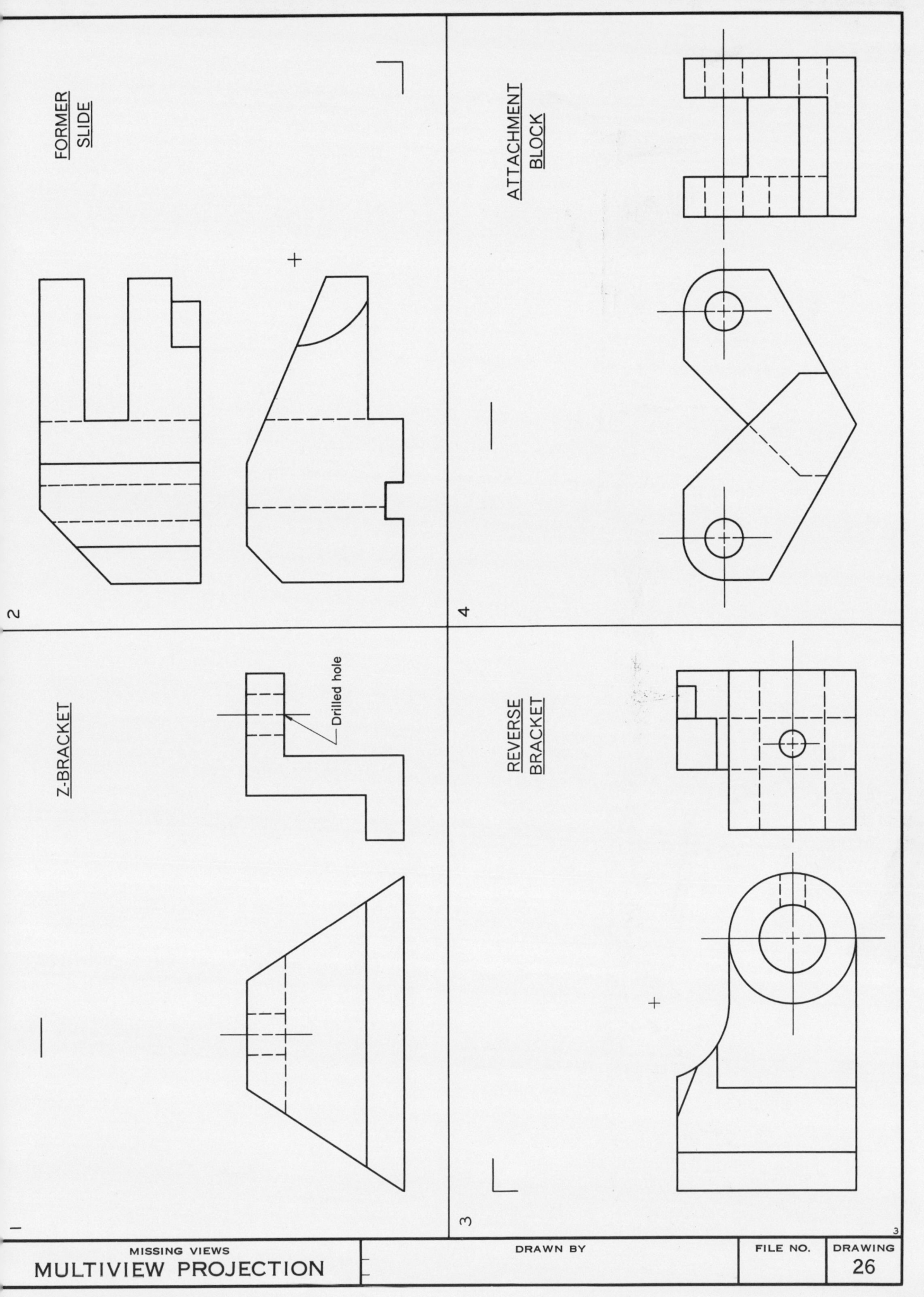

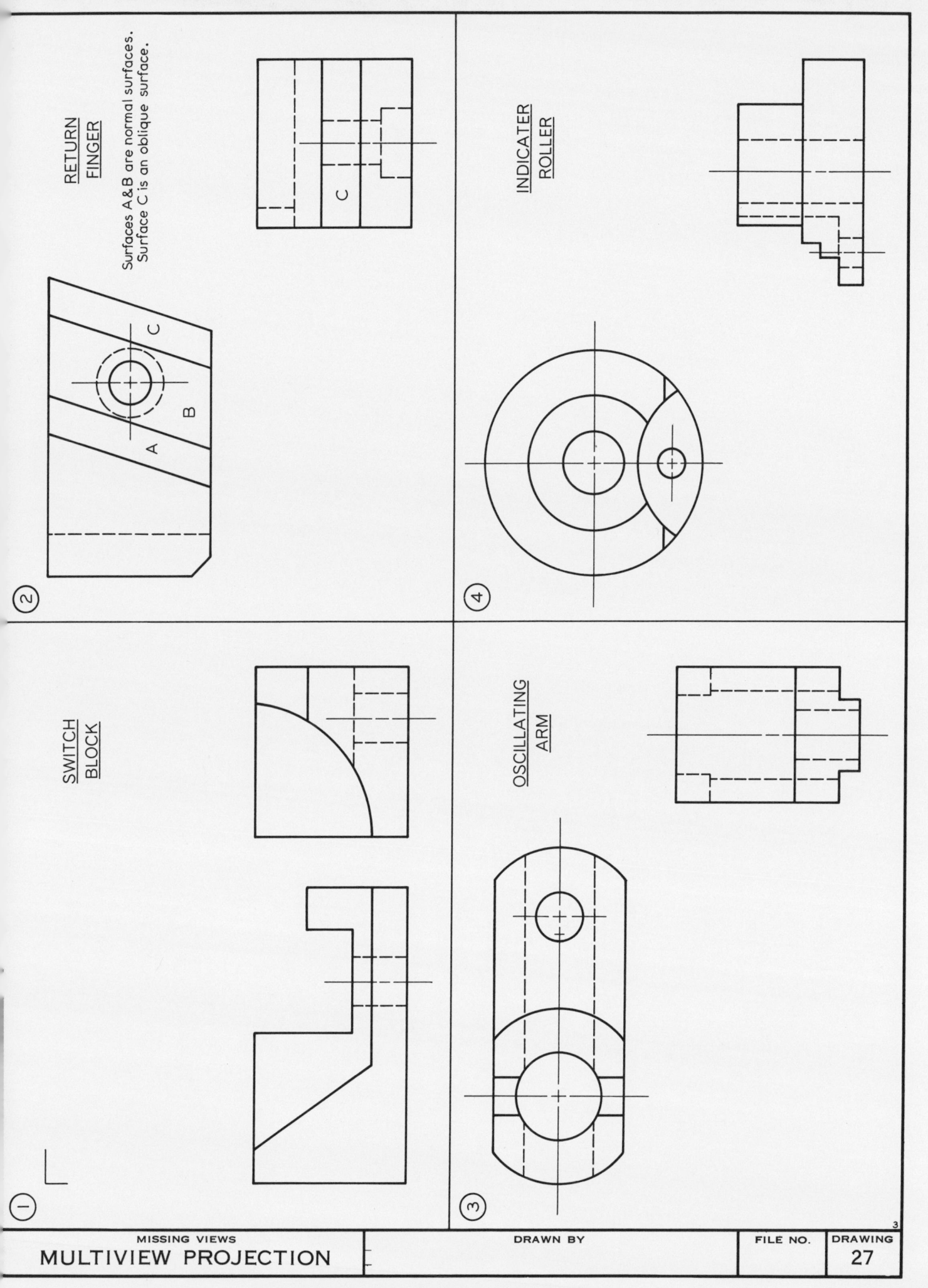

1

HOLDER CAM

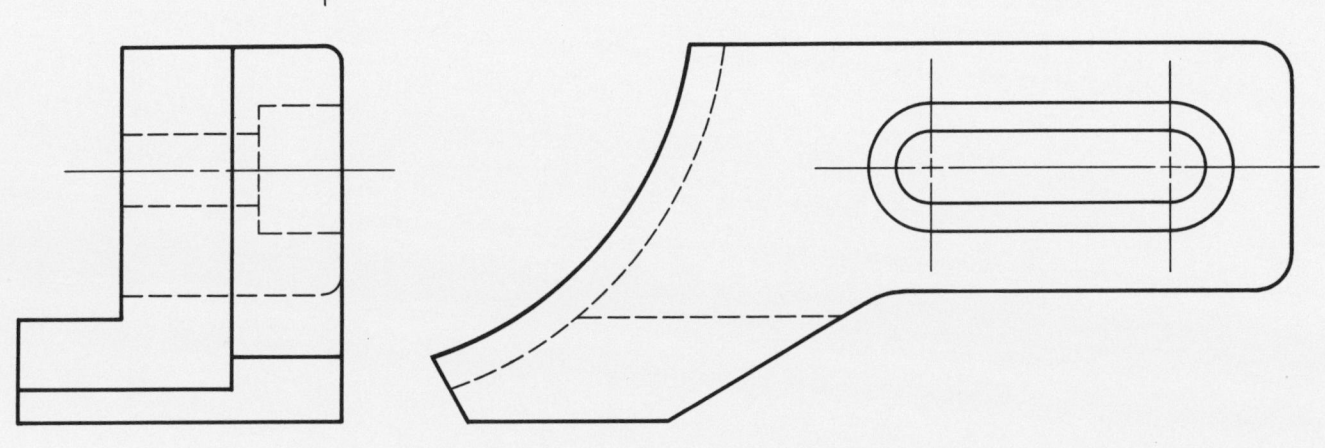

2

AUTOMATIC GUIDE

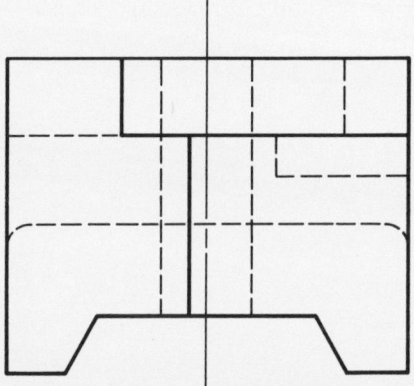

| MISSING VIEWS | DRAWN BY | FILE NO. | DRAWING |
| MULTIVIEW PROJECTION | | | 29 |

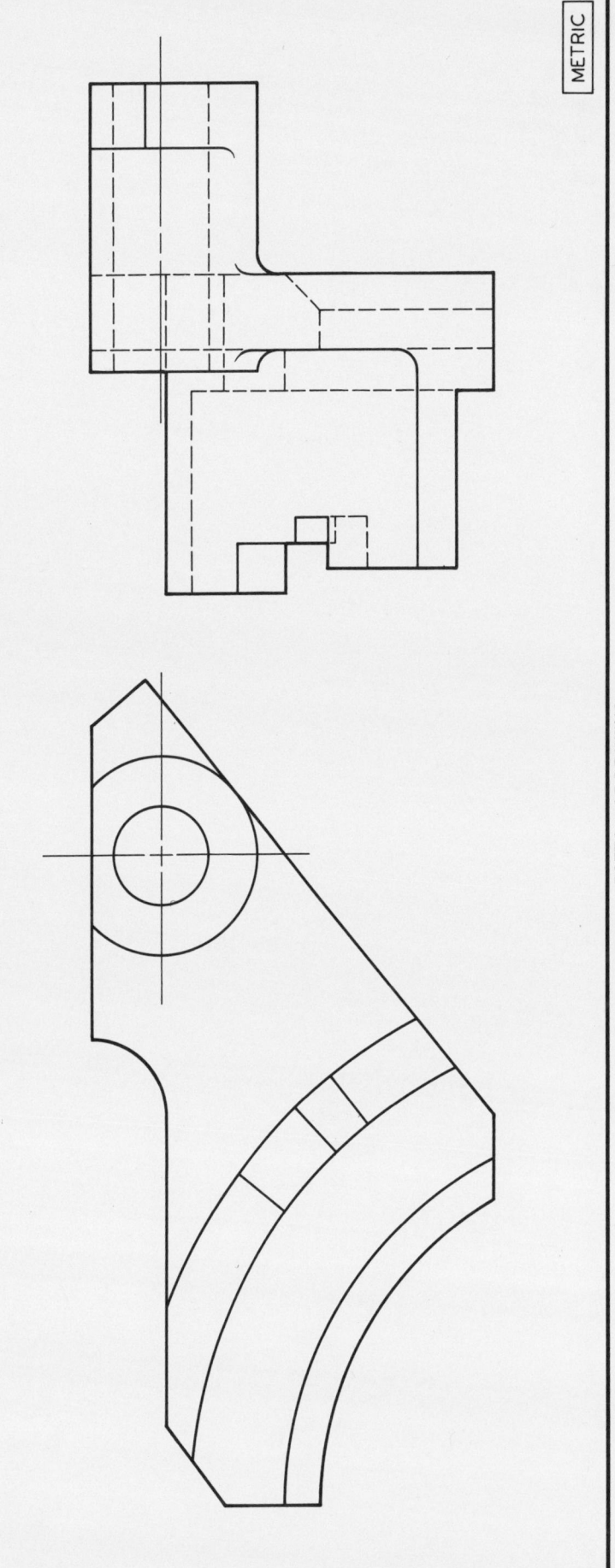

CONTROL LEVER
FOR AUTOMATIC MIXER

Draw top view.
Fillets & rounds 3R

BOTTOM OF DRAWING

METRIC

MISSING VIEW
MULTIVIEW PROJECTION

DRAWN BY

FILE NO.

DRAWING
30

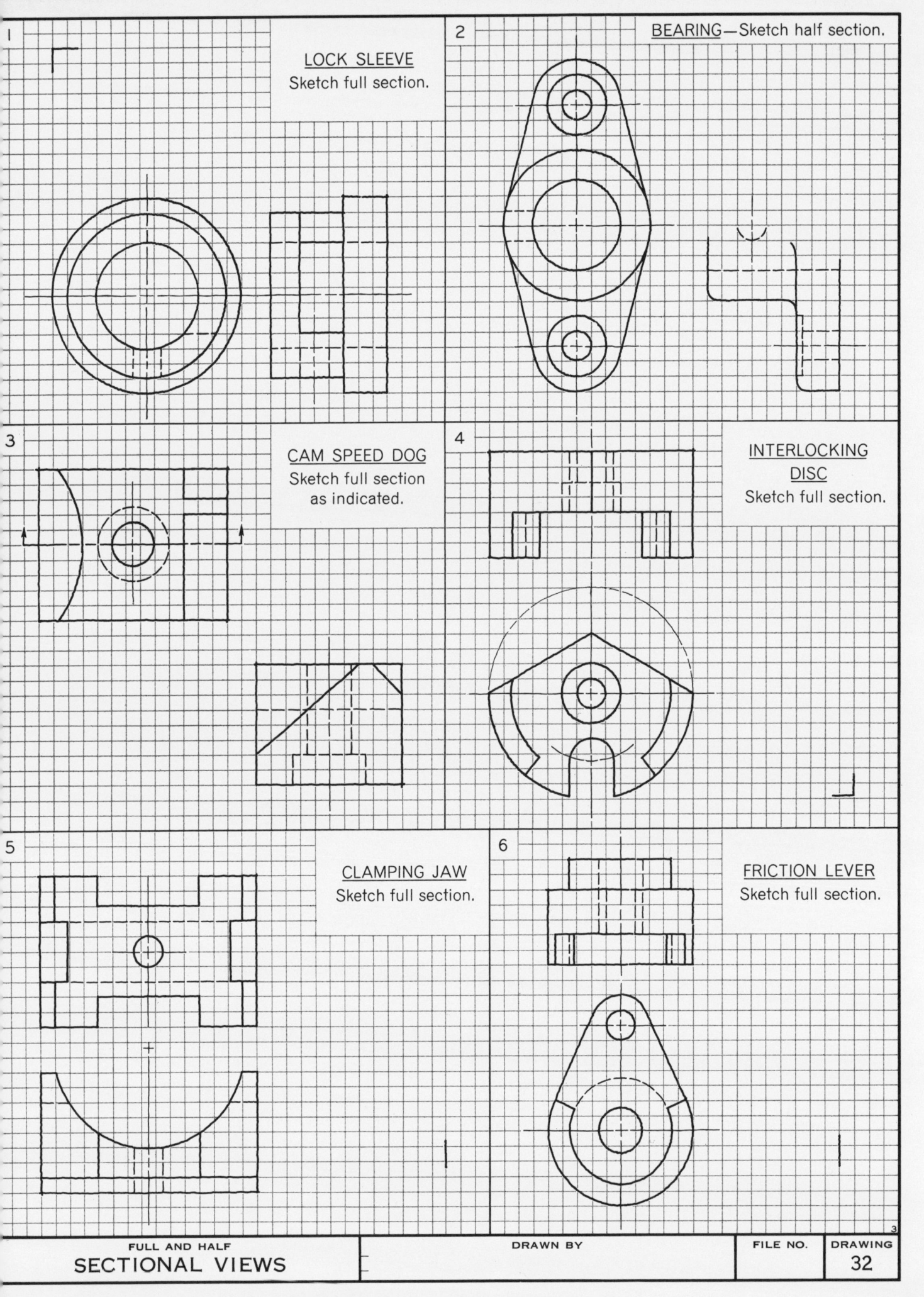

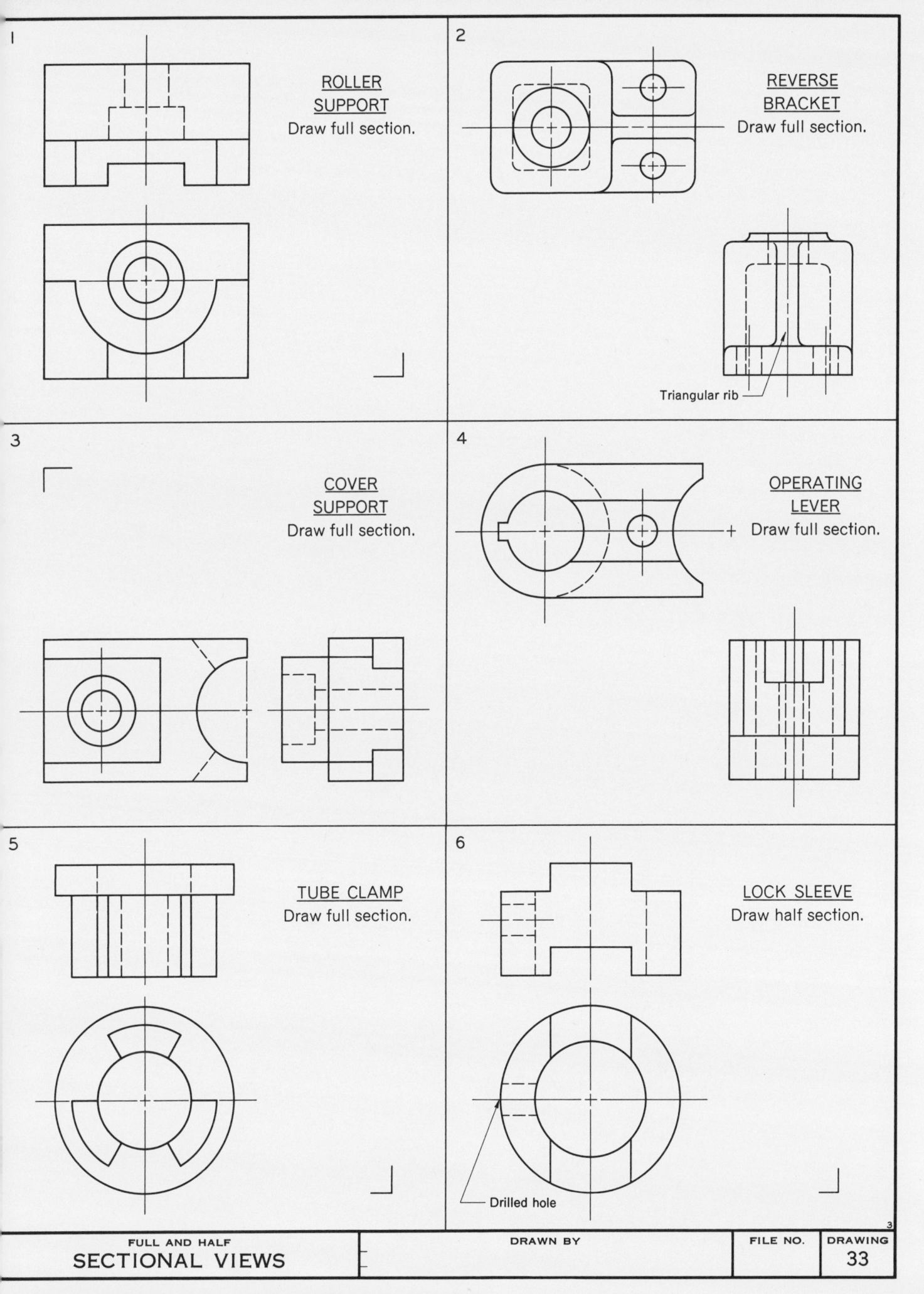

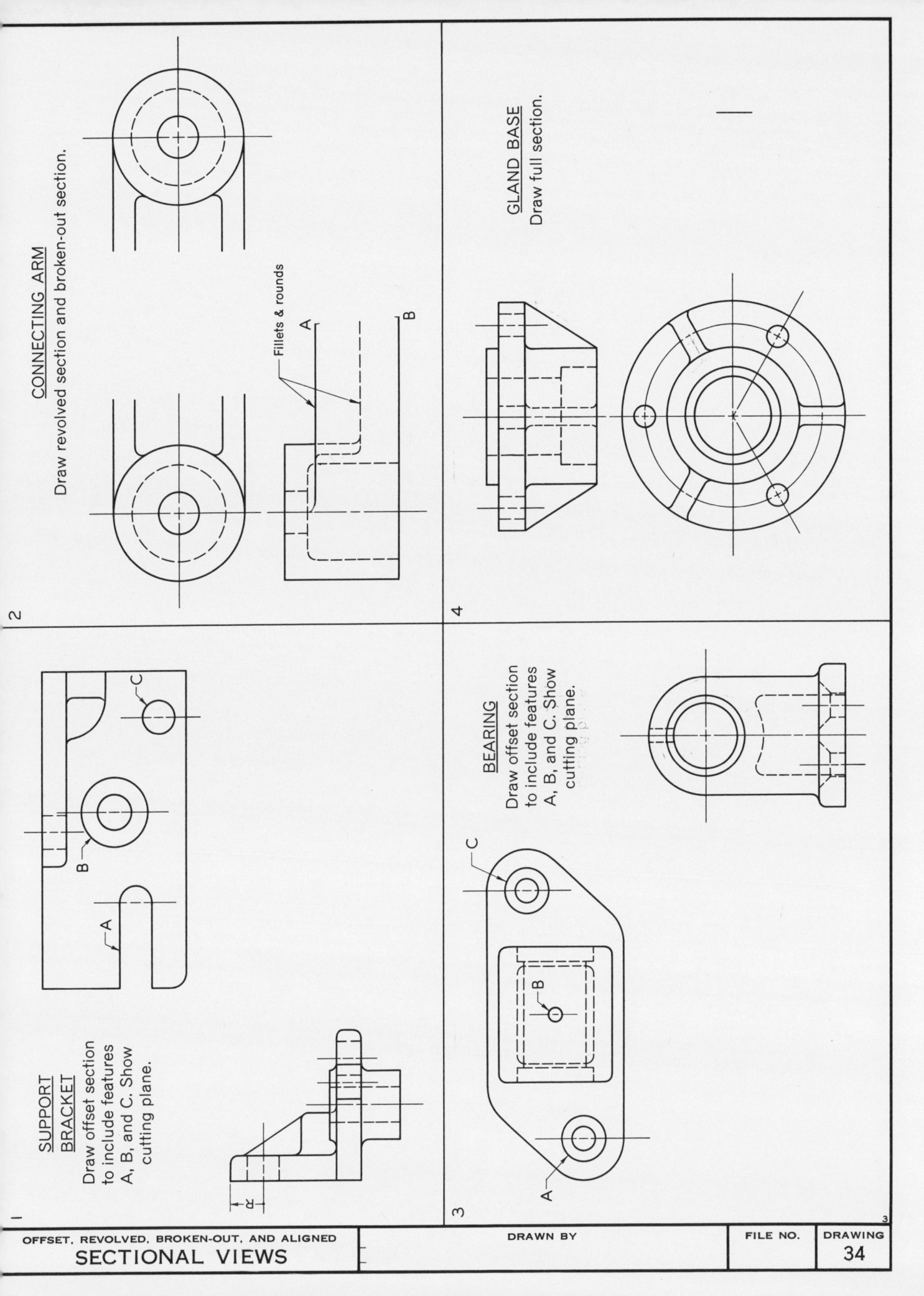

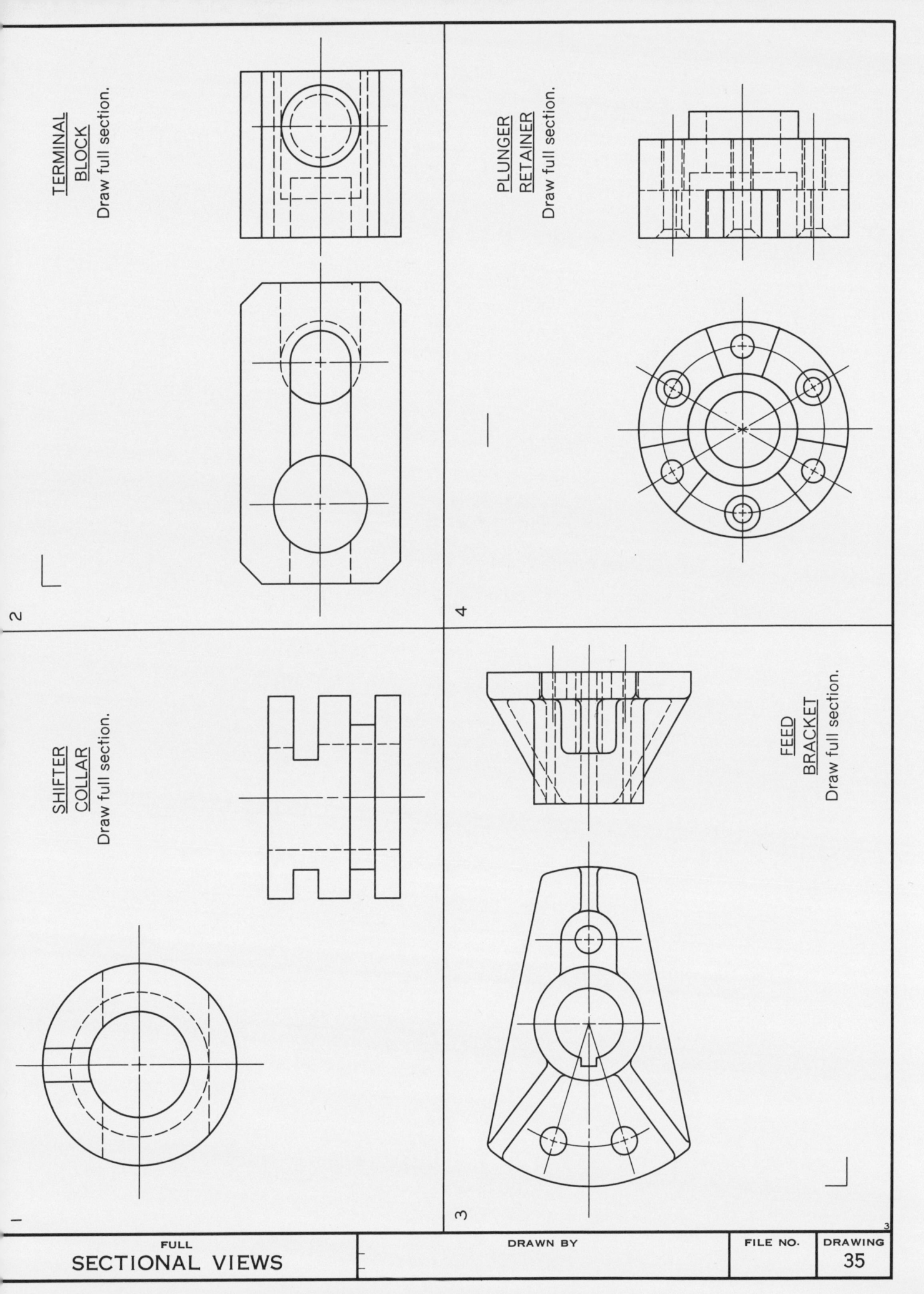

(1)

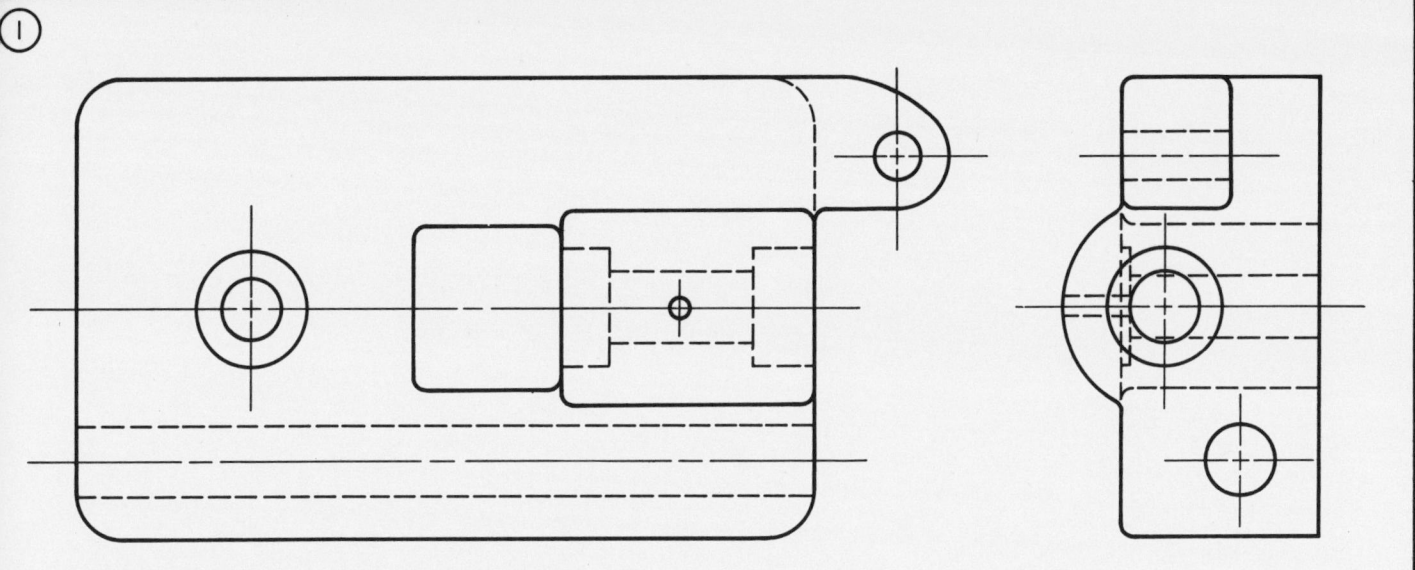

GUIDE SHOE

Draw full section.

(2)

SPECIAL VALVE

Draw removed sections.

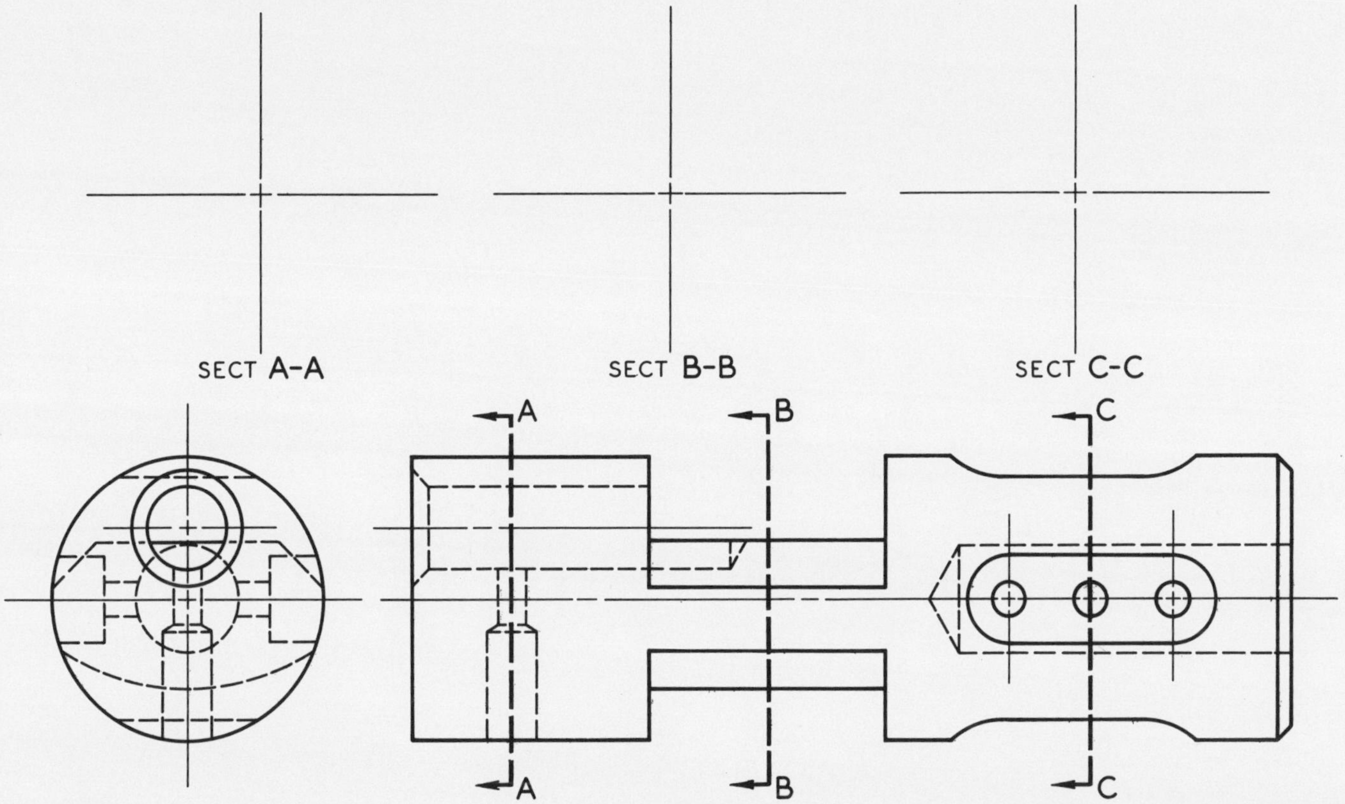

SECT A-A
SECT B-B
SECT C-C
A
B
C
A
B
C

FULL AND REMOVED SECTIONAL VIEWS	DRAWN BY	FILE NO.	DRAWING 37

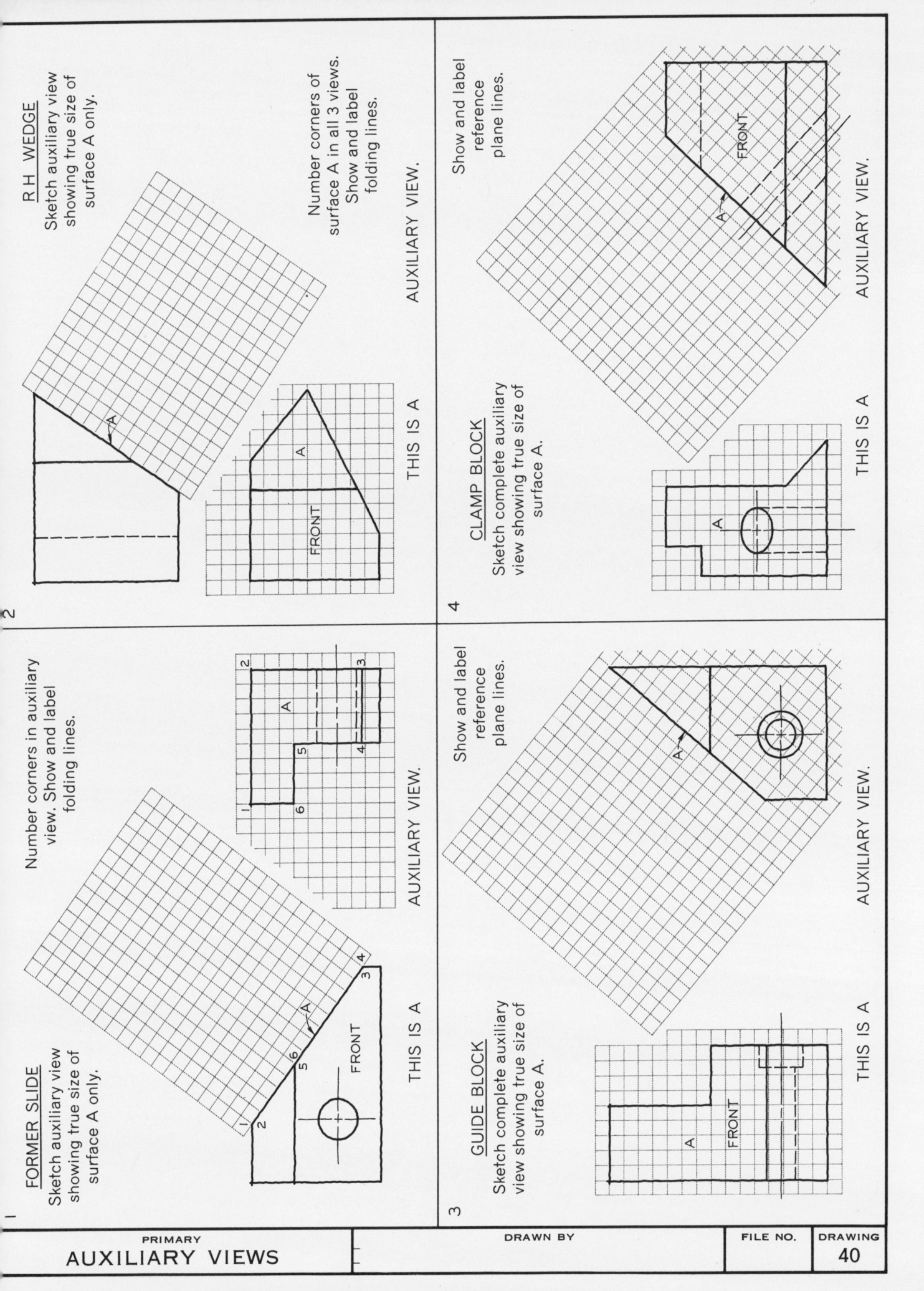

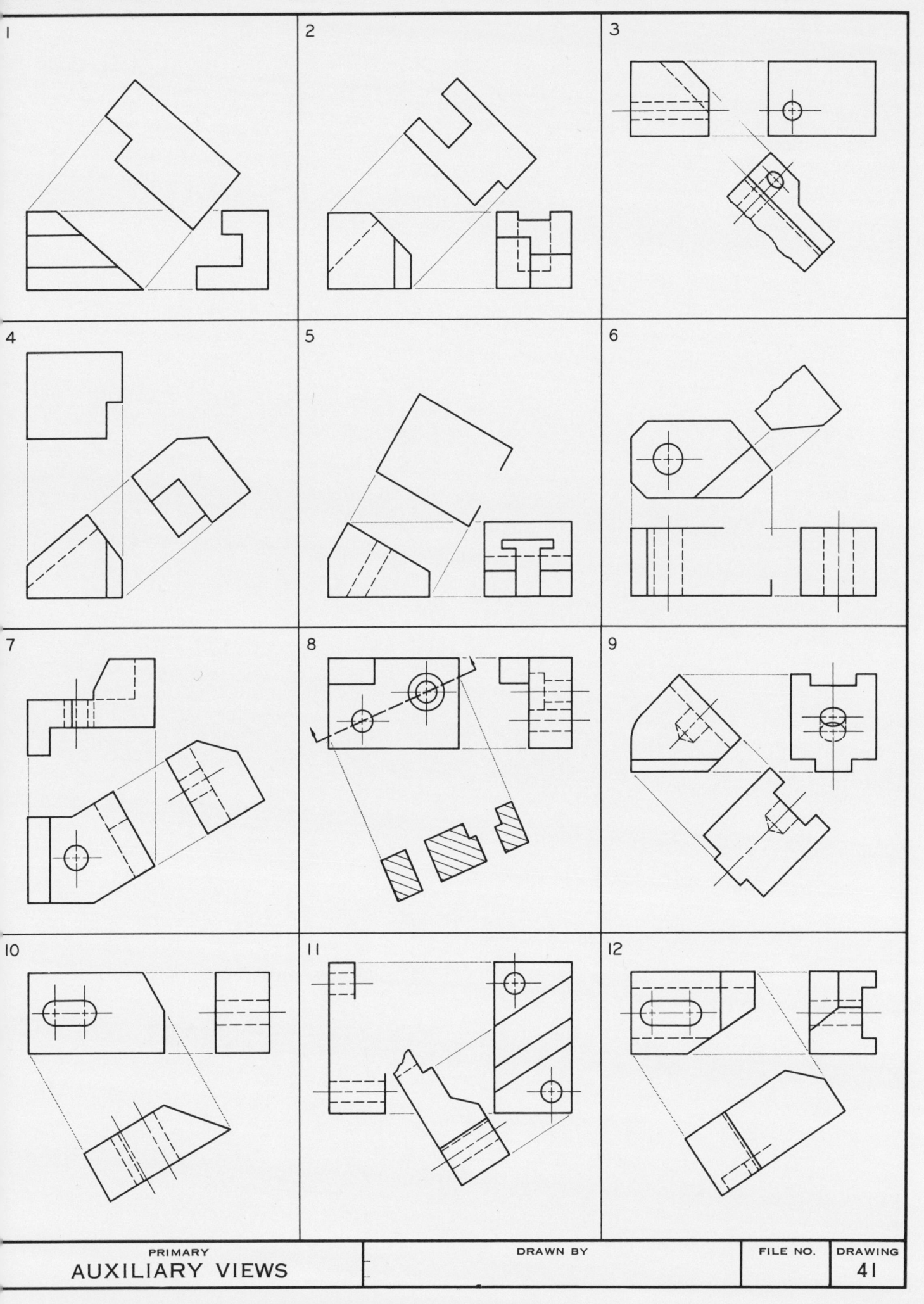

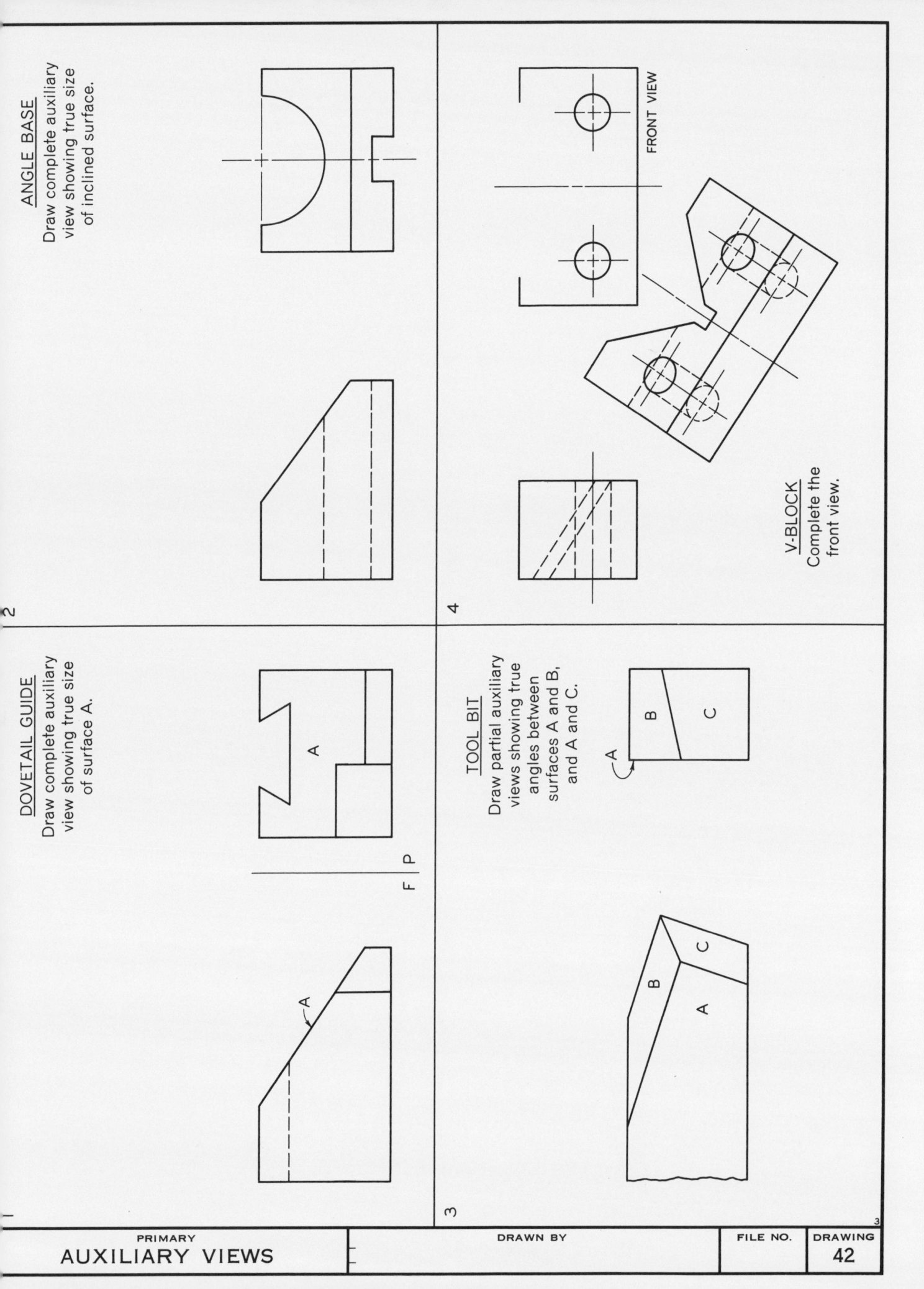

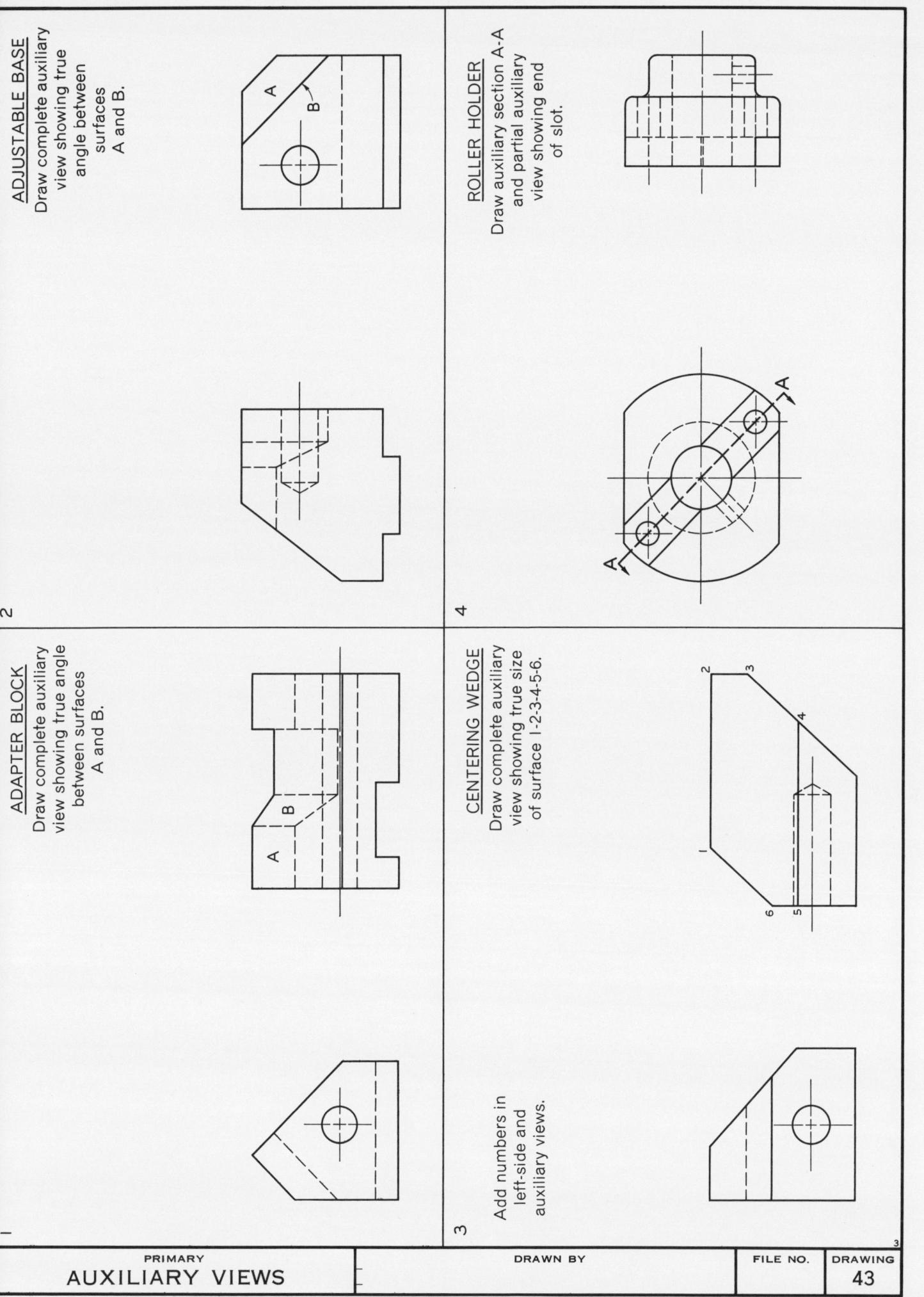

1

ATTACHMENT CAM
Draw complete auxiliary view showing angle between surfaces A and B; then draw right-side view.

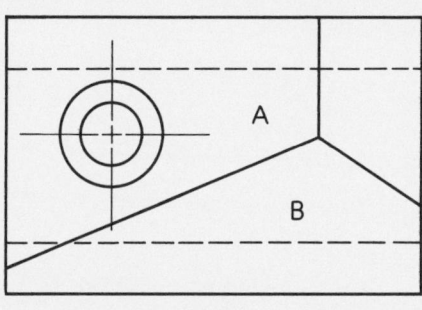
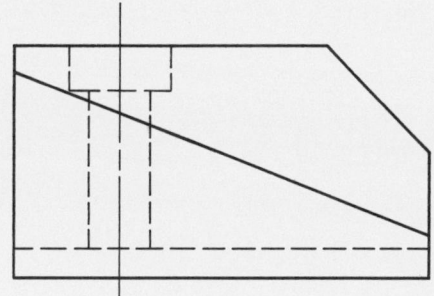

2

DOVETAIL SLIDE
Draw complete auxiliary view showing angle of dovetail; then draw top view.

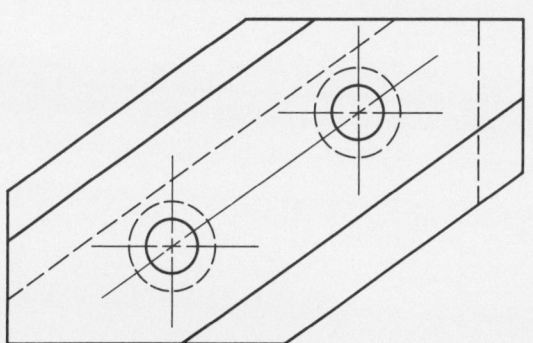
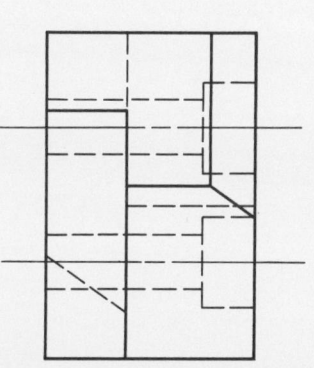

| PRIMARY AUXILIARY VIEWS | DRAWN BY | FILE NO. | DRAWING 44 |

1

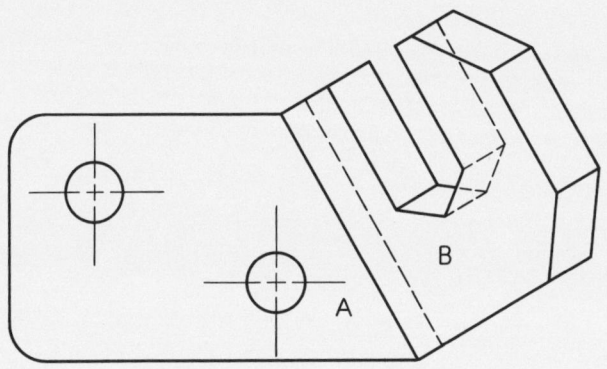

ANGLE BRACKET
Draw complete primary auxiliary view showing 135° angle between surfaces A and B; then draw partial secondary auxiliary view showing true size of member B. Thickness of members–10mm.

2

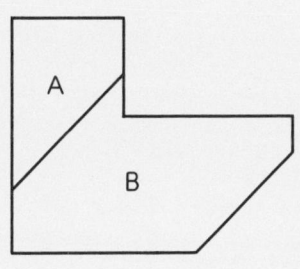

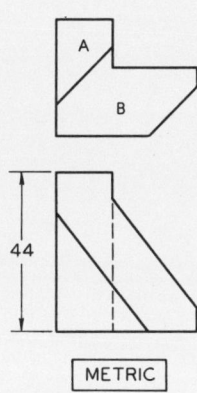

METRIC

STOP BLOCK
Draw complete primary auxiliary view showing angle between surfaces A and B; then draw complete secondary auxiliary view showing true size of surface B.

SECONDARY AUXILIARY VIEWS | DRAWN BY | FILE NO. | DRAWING 46

1

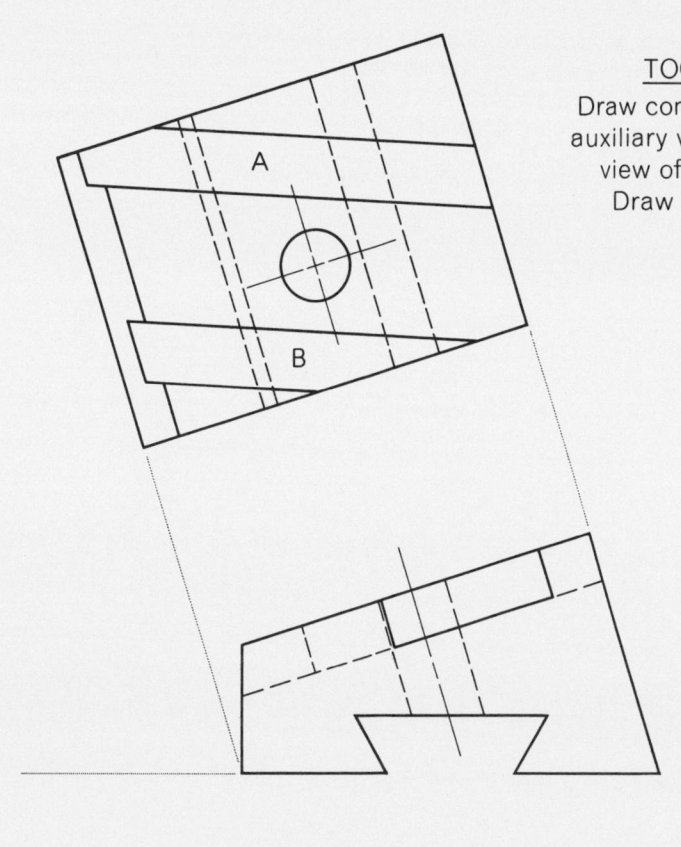

TOOL BLOCK
Draw complete secondary auxiliary view showing end view of slots A and B. Draw left-side view.

2

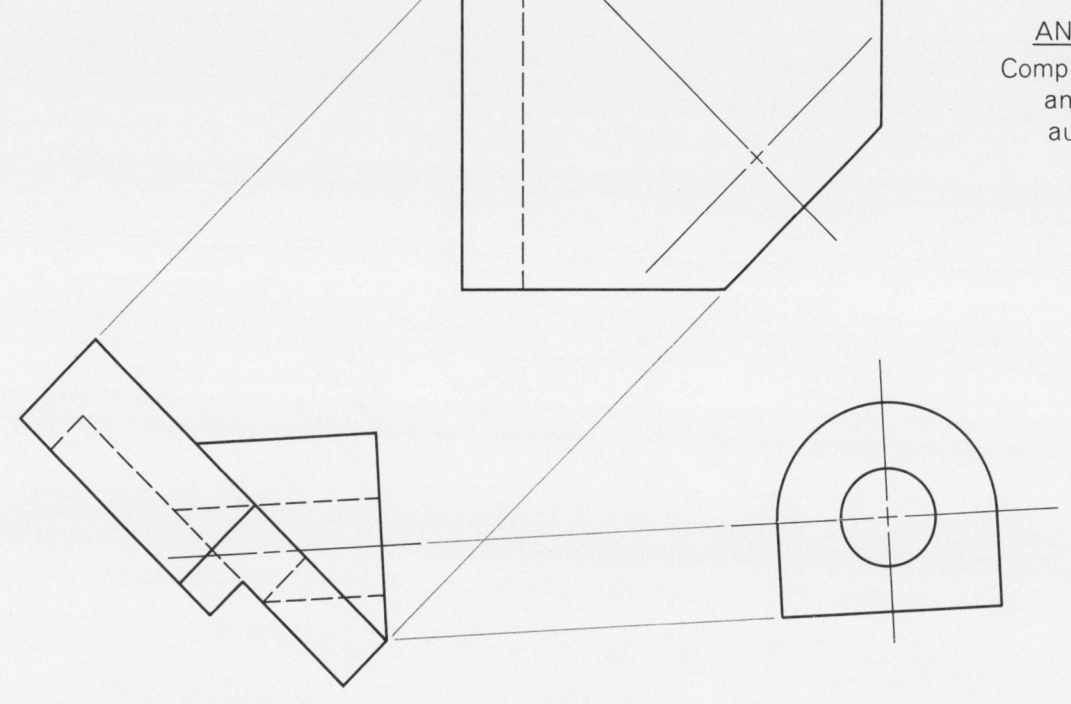

ANCHOR BASE
Complete the top view and secondary auxiliary view.

| SECONDARY AUXILIARY VIEWS | DRAWN BY | FILE NO. | DRAWING 47 |

1. LEVER BRACKET
Draw front and top views with object revolved so semi-circular portion shows true size in front view.

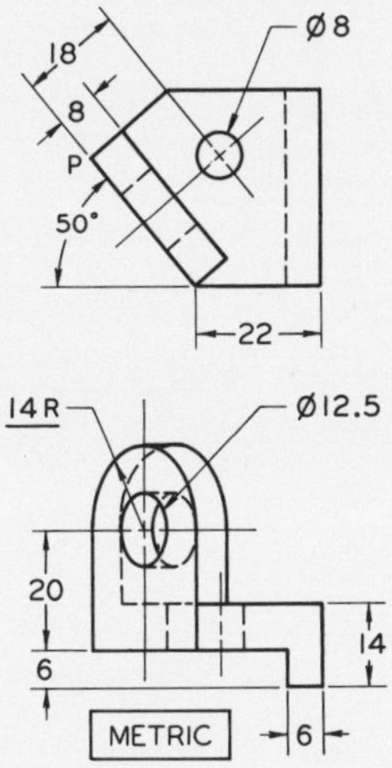

2. WEDGE
Draw front and top views with object revolved so angle between surfaces A and B shows true size in front view. Dimension the angle.

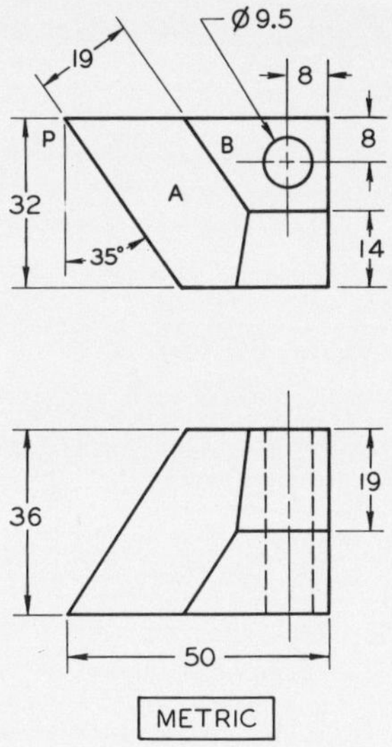

3. ELEVATION SLIDE
Draw front view revolved until angle between surfaces A and B is true size in right-side view. Draw right-side and top views, and dimension angle between A and B

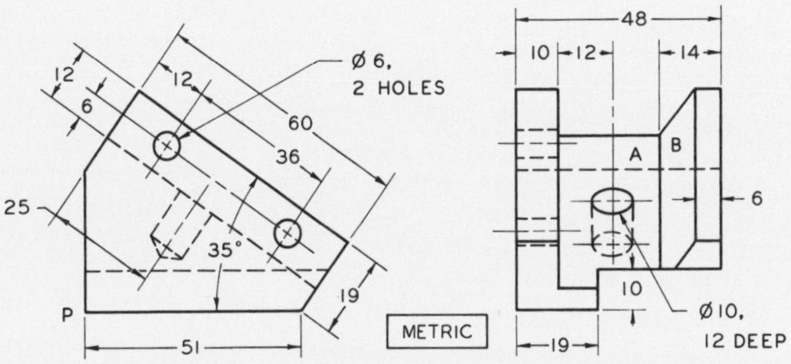

PRIMARY REVOLUTIONS

DRAWING 49

REGULATOR BLOCK

Revolve surface A until it appears true size in the top view. Revolve surface B until it appears true size in the front view.

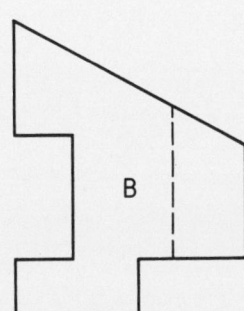

2. BEVEL CLAMP

Find true size of surface A by means of an auxiliary view showing the edge view of the surface; then by revolution obtain true size in front view.

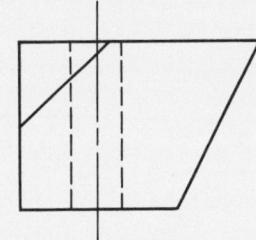

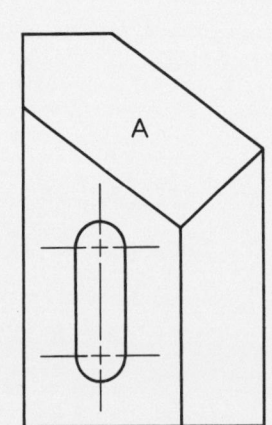

3. TRIP LEVER

Draw Section A-A showing inclined arm revolved.

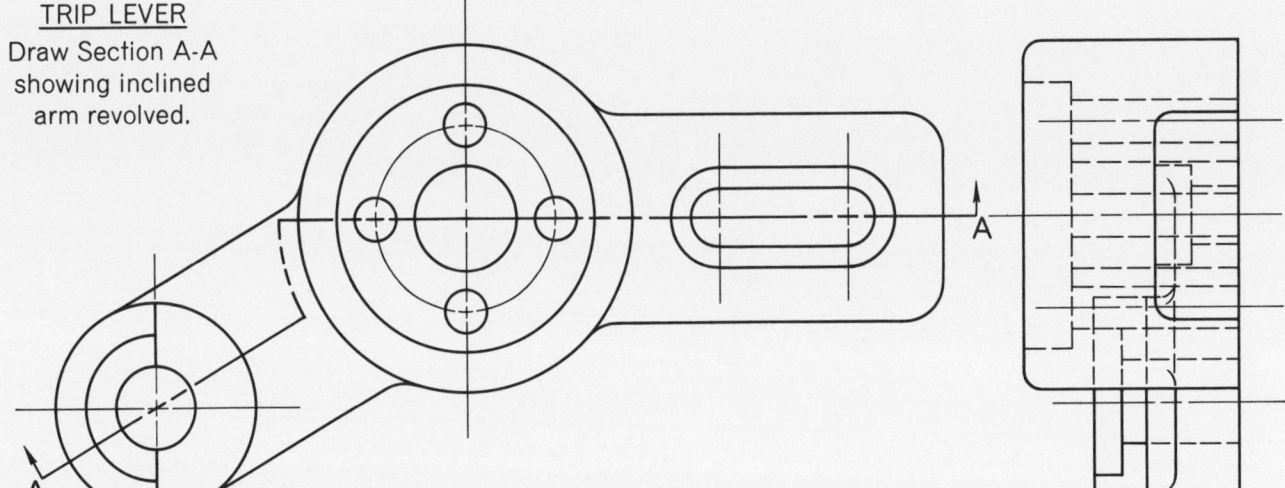

PRIMARY REVOLUTIONS — DRAWING 50

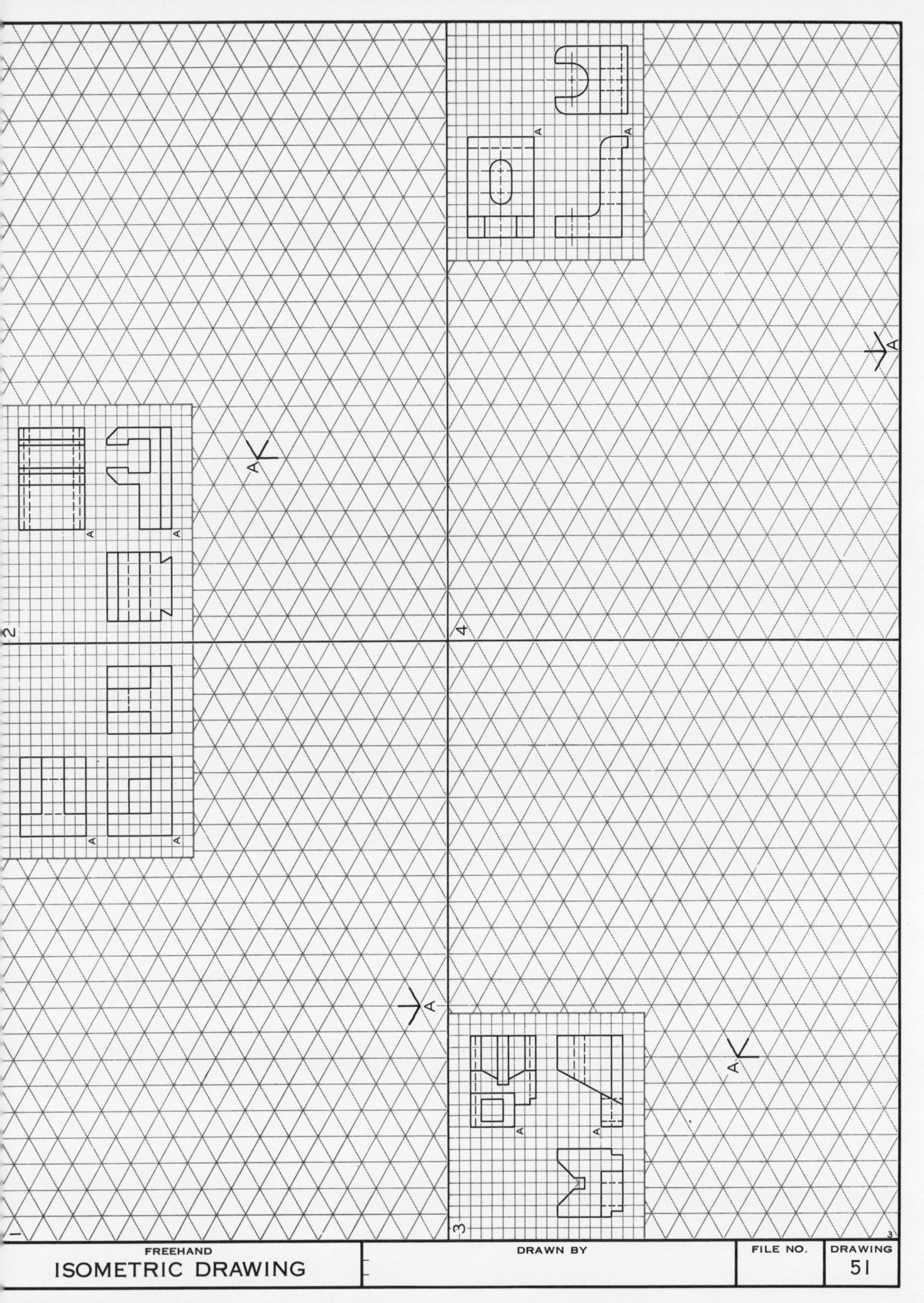

FREEHAND
ISOMETRIC DRAWING

DRAWN BY

FILE NO.

DRAWING 51

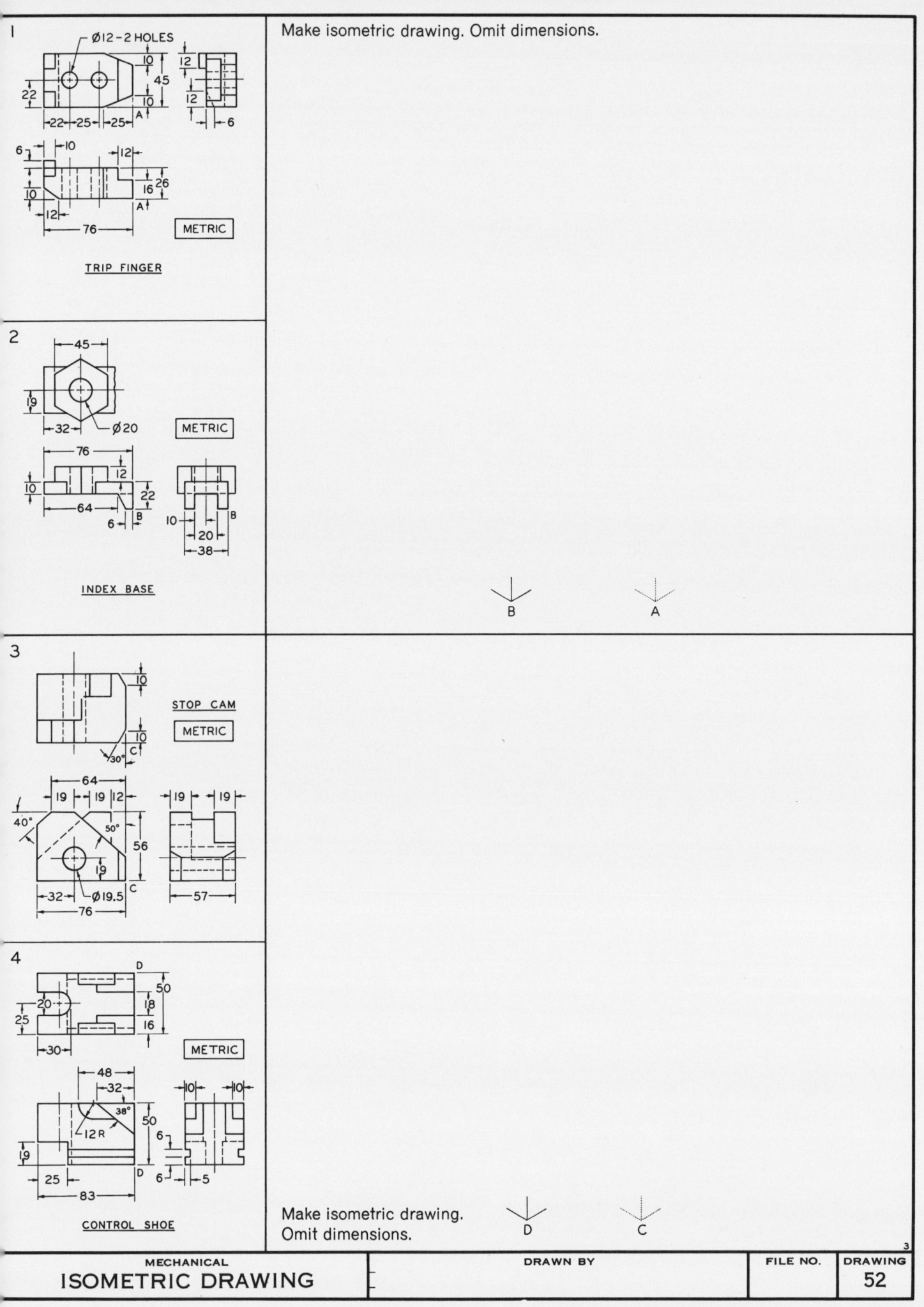

1

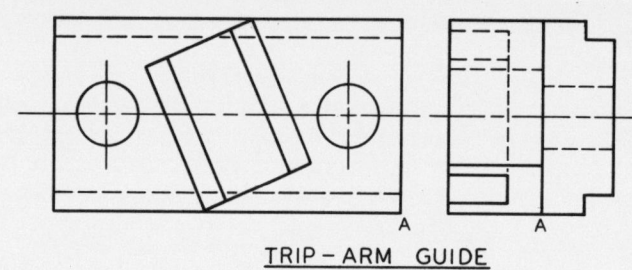

TRIP – ARM GUIDE
HALF SIZE

Make full-size isometric drawing.

2

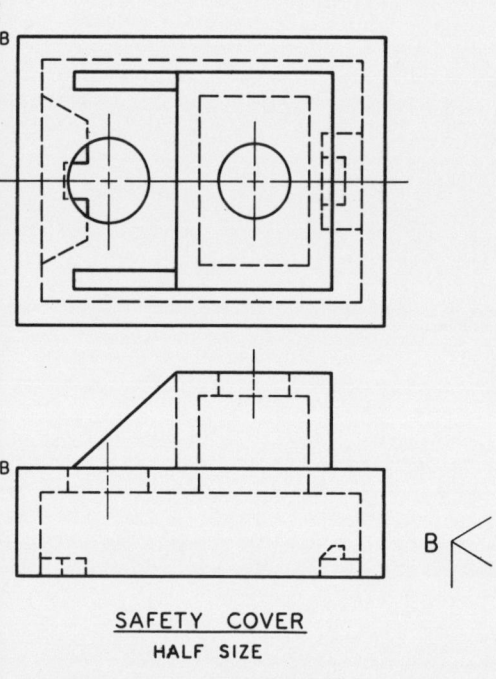

SAFETY COVER
HALF SIZE

Make full-size full-section isometric drawing.

| MECHANICAL ISOMETRIC DRAWING | DRAWN BY | FILE NO. | DRAWING 53 |

1

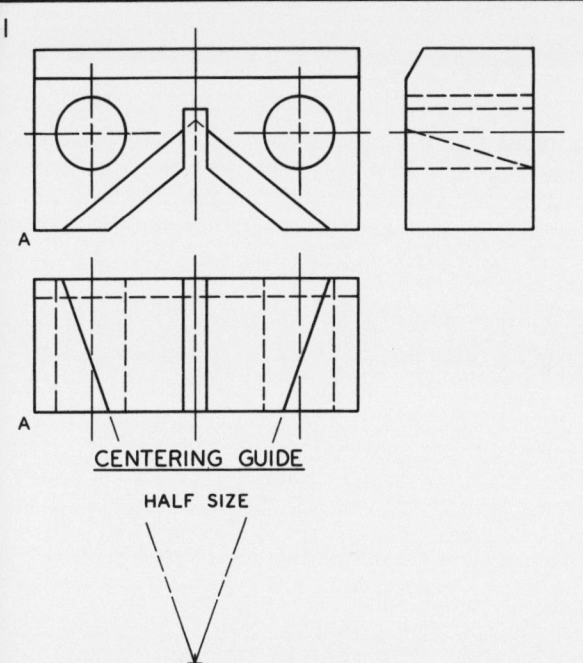

CENTERING GUIDE
HALF SIZE

Make full-size isometric drawing.

2

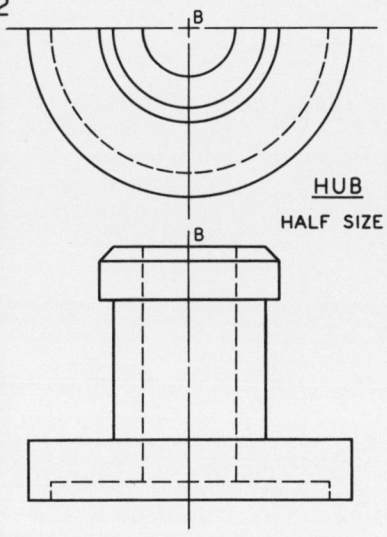

HUB
HALF SIZE

Make full-size isometric drawing in
half-section or full-section as assigned.

| MECHANICAL ISOMETRIC DRAWING | DRAWN BY | FILE NO. | DRAWING 54 |

1

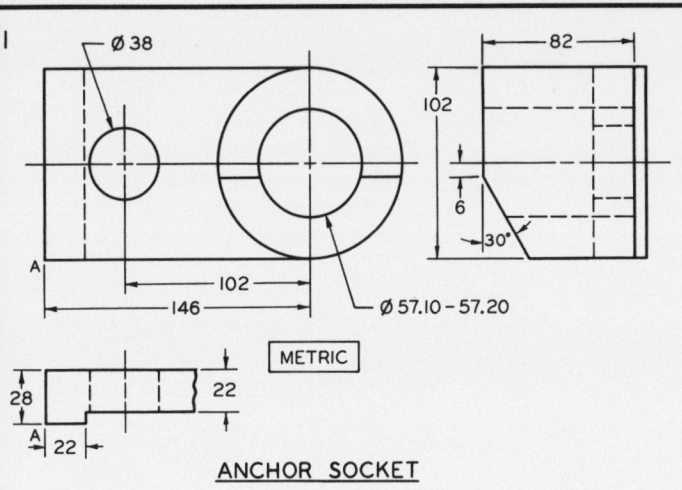

ANCHOR SOCKET

Make half-size isometric drawing.

A

2

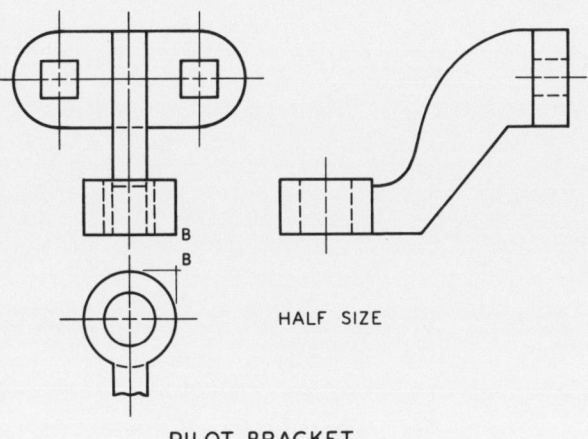

HALF SIZE

PILOT BRACKET

Make full-size isometric drawing.

B

MECHANICAL ISOMETRIC DRAWING

DRAWING 55

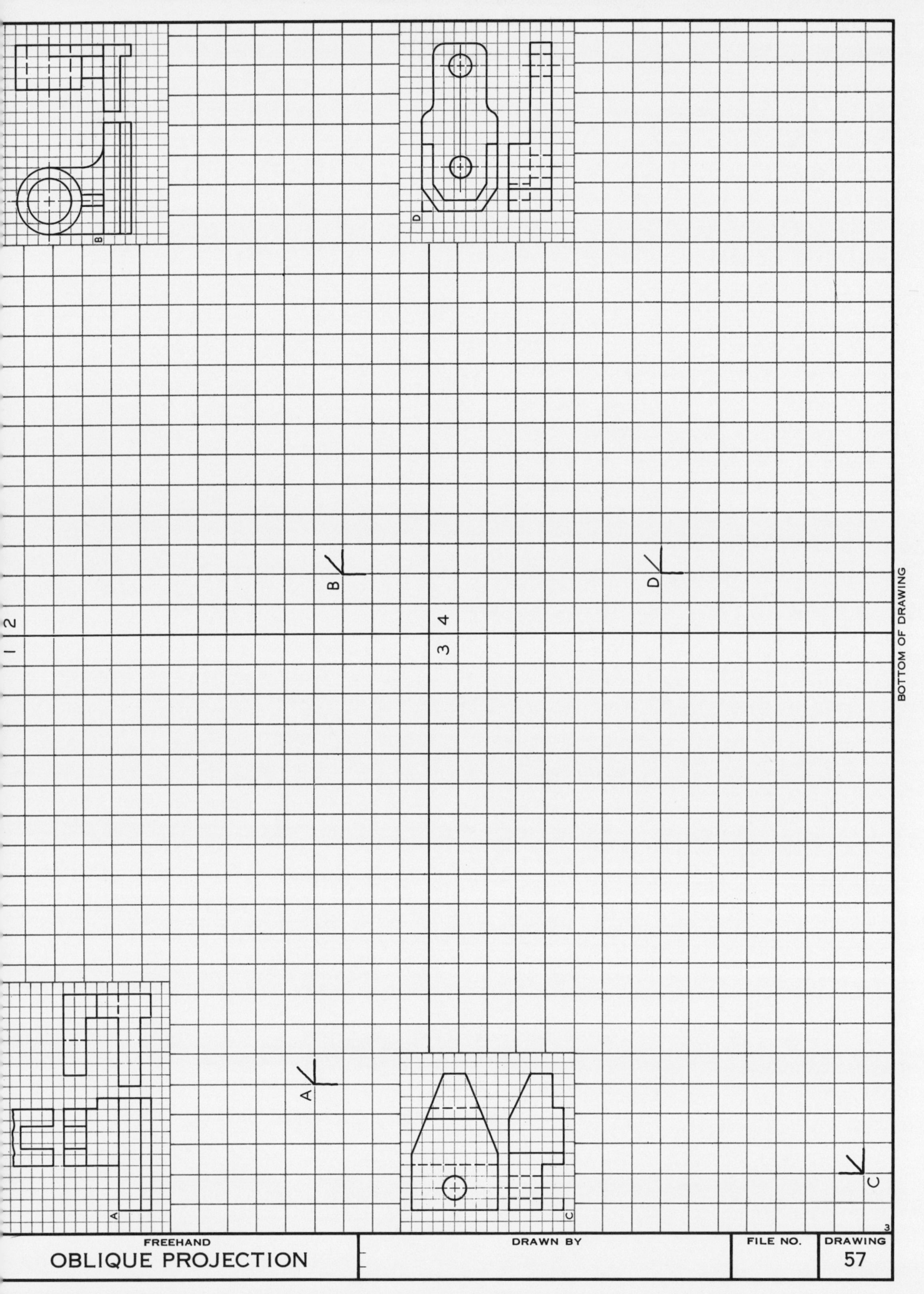

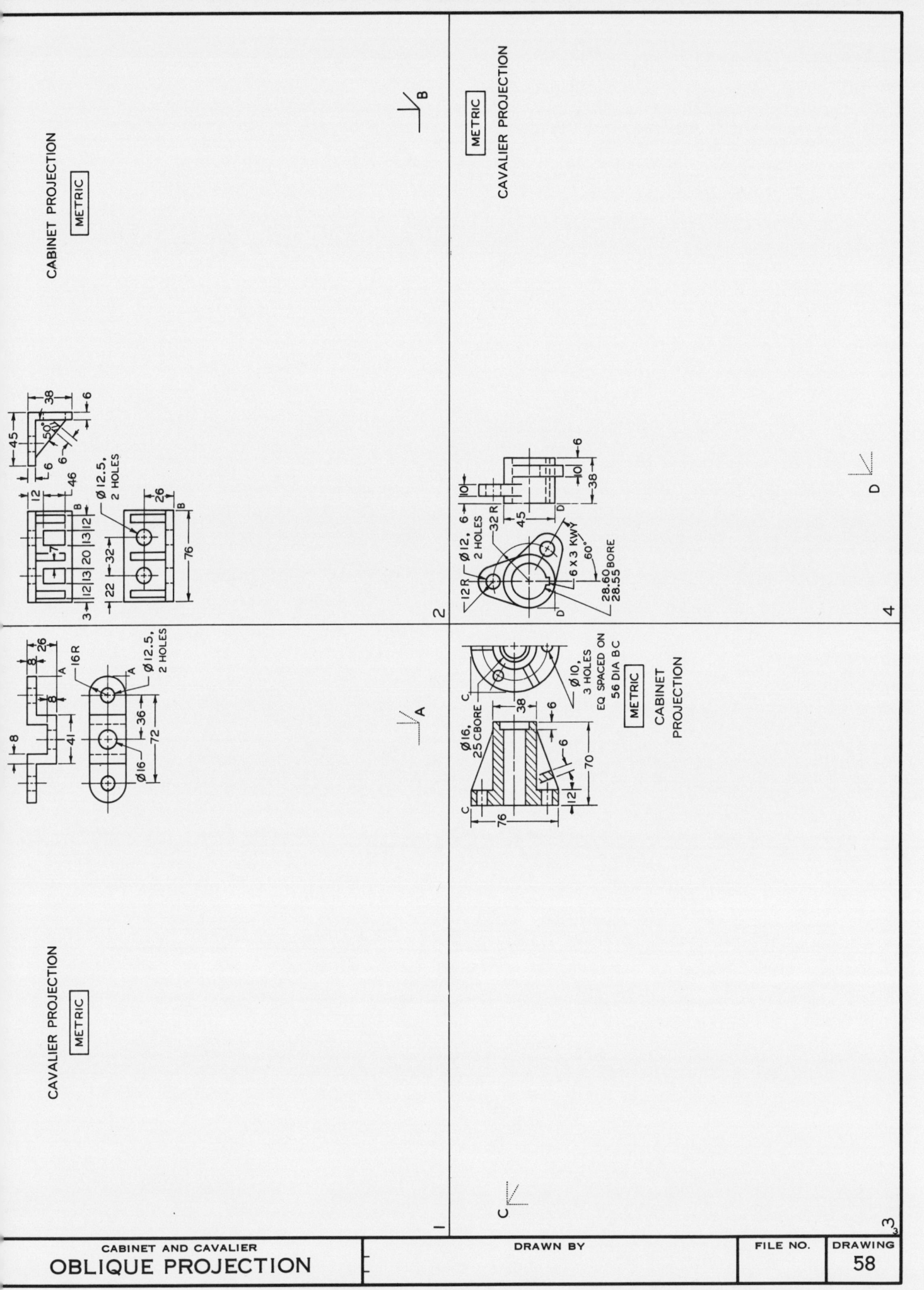

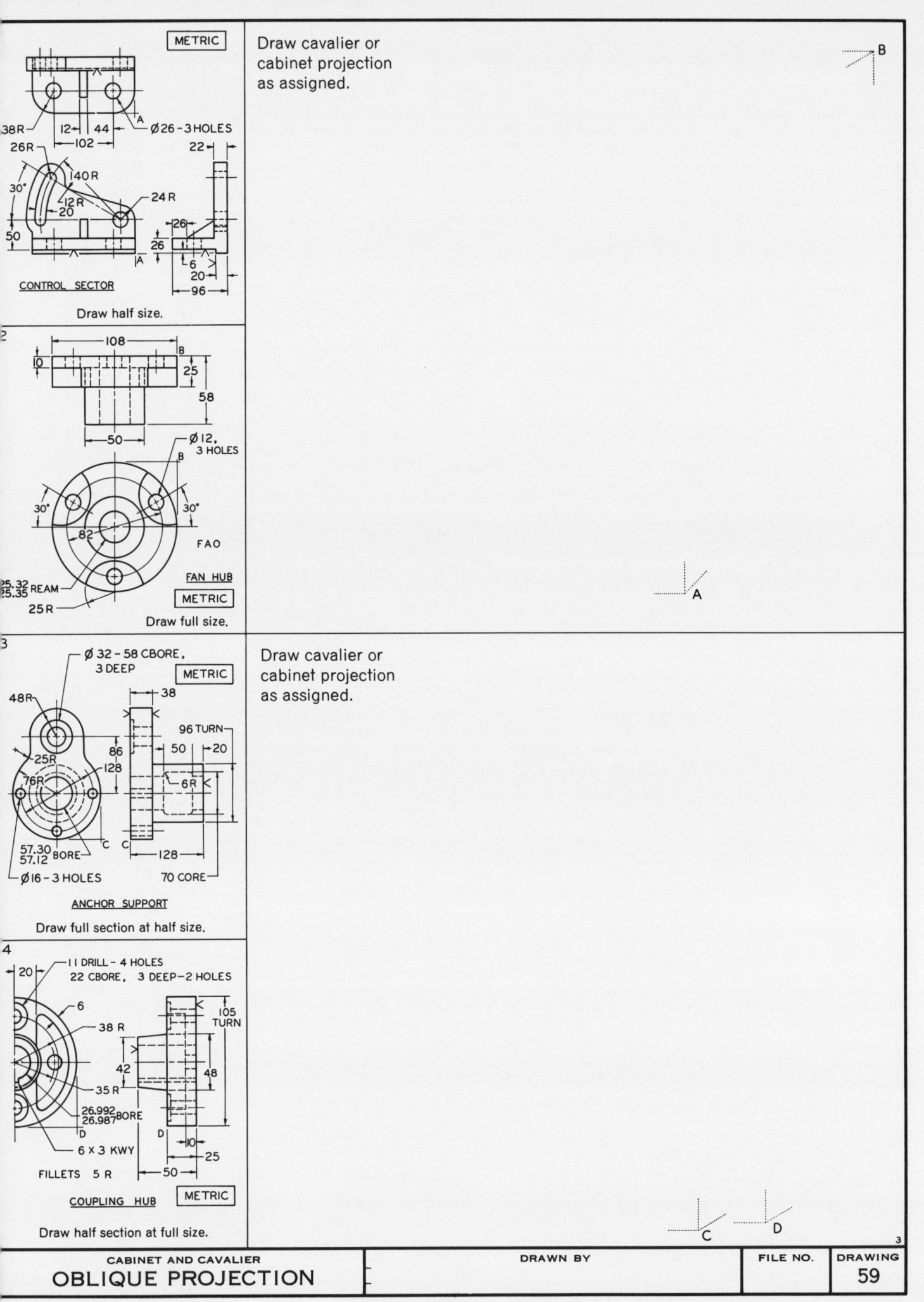

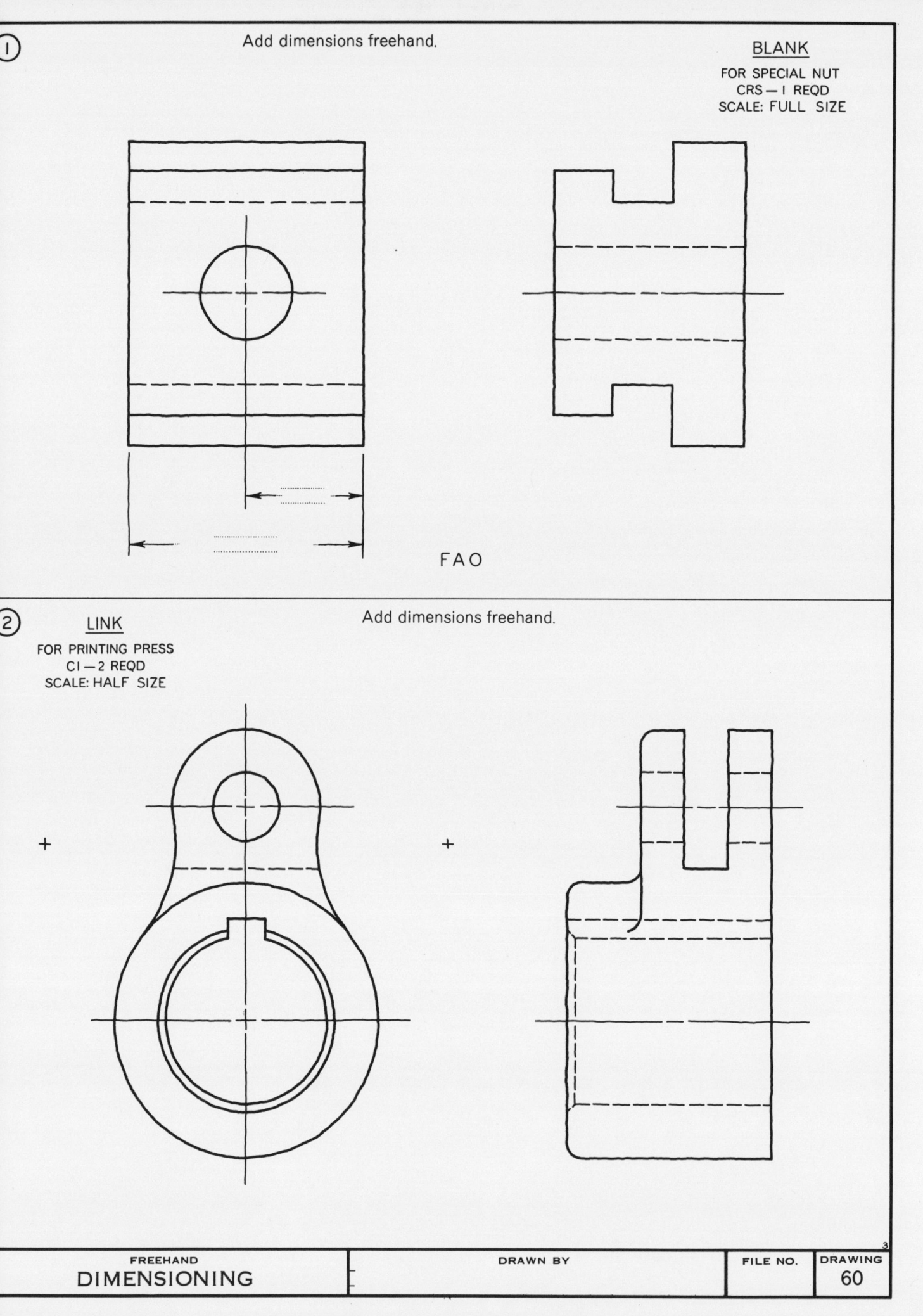

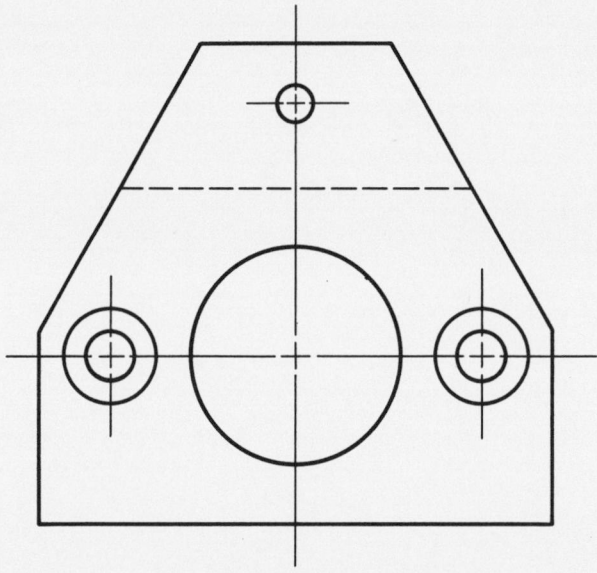

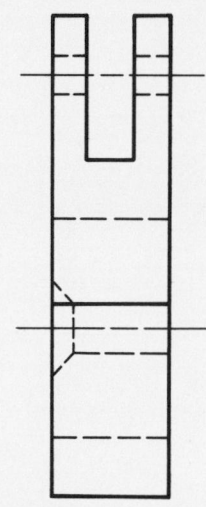

CARRIER
FOR RADIAL BEARING
CRS — 4 REQD
FULL SIZE

Add dimensions mechanically.

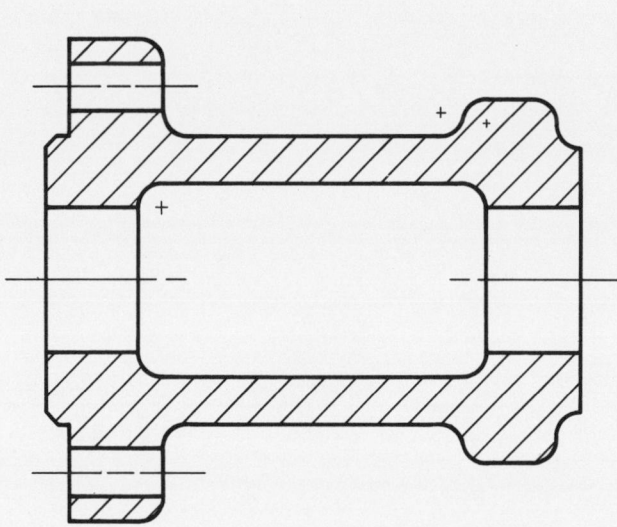

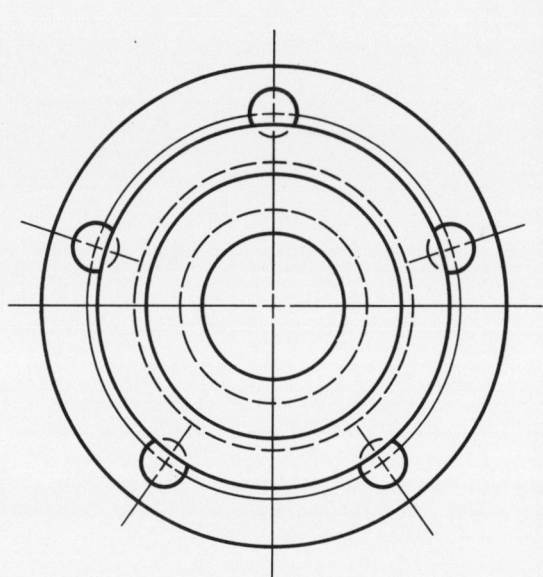

SUPPORT FRAME
FOR PRESS
C1 — 1 REQD
FULL SIZE

Add dimensions mechanically.

MECHANICAL DIMENSIONING   DRAWN BY   FILE NO.   DRAWING 61

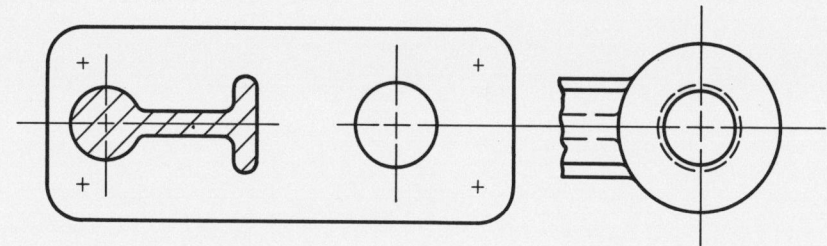

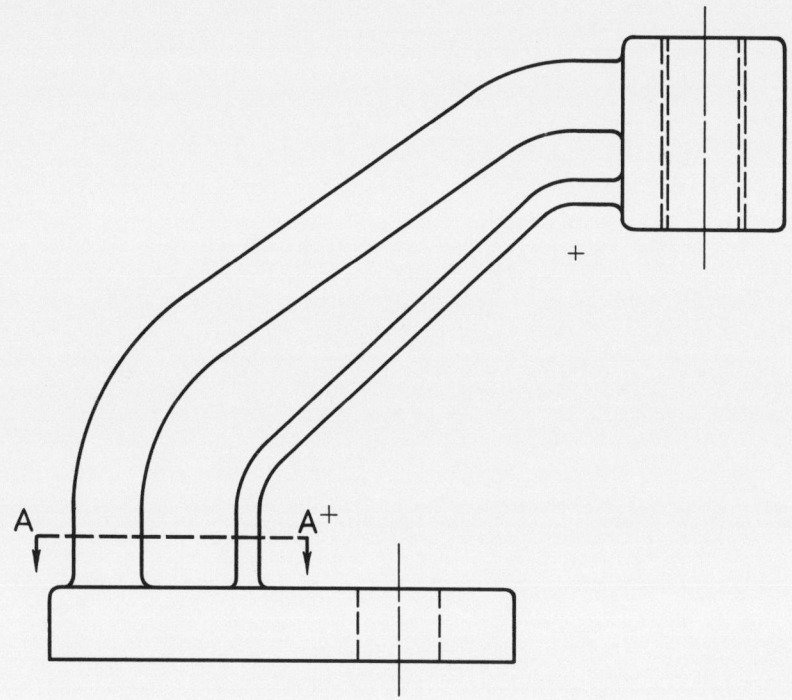

T-SLOT CLAMP

UNSPECIFIED FILLETS & ROUNDS .06R   FULL SIZE

(1) FRAME
CAST STEEL
1 REQD

MATING PARTS
DIMENSIONING

DRAWING 63

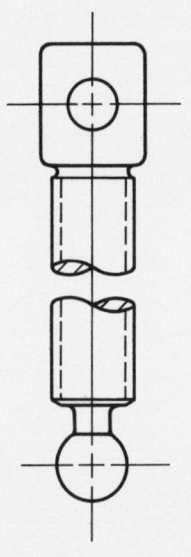

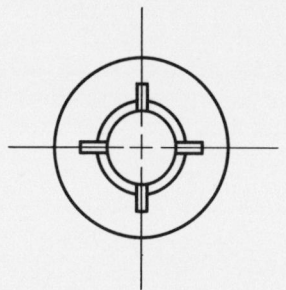

③ <u>PAD</u>   CRS - I REQD

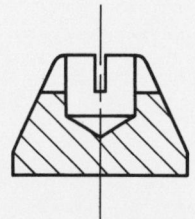

② <u>CLAMP SCREW</u>
CRS - I REQD

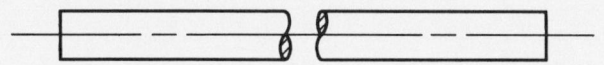

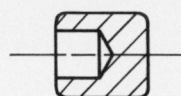

⑤ <u>HANDLE CAP</u>
CRS - 2 REQD

STANDARD PARTS

⑥ I - MI0 x 1.5
HEAVY HEX NUT

⑦ I - 12 x 22 x 4.5 T-SLOT
WASHER

⑧ I - MI0 x 1.5 - 50 LONG
T-SLOT BOLT

④ <u>HANDLE</u>   CRS - I REQD

UNSPECIFIED FILLETS & ROUNDS 1.5R   FULL SIZE

| MATING PARTS DIMENSIONING | DRAWN BY | FILE NO. | DRAWING 64 |

This sheet and the following sheet contain the views of the parts, a standard parts list, and an assembly pictorial of the Roller Guide. Add dimensions, notes, etc., to complete the details. Use RC 6 limits for large hole in base.

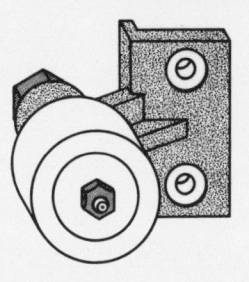

① BASE
FOR
ROLLER GUIDE
CAST IRON
1 REQD
SCALE: 1 = 1

SECTION A-A

FILLETS AND ROUNDS 1.5 R

BOTTOM OF DRAWING

METRIC

MATING PARTS
DIMENSIONING

DRAWN BY

FILE NO.

DRAWING
66

STANDARD PARTS

⑤ 1 – M16 × 2 HEXAGON NUT
⑥ 1 – 11/16 – AMERICAN NATIONAL STD REGULAR LOCKWASHER
⑦ 1 – NO. 8585 HYDRAULIC GREASE FITTING
⑧ 1 – NO. 405 WOODRUFF KEY

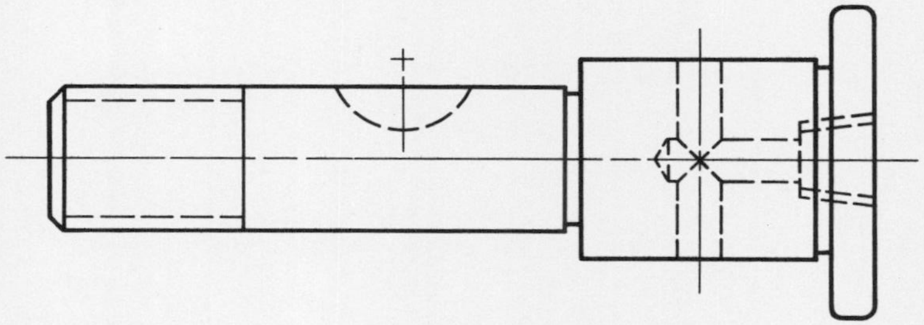

SCHEDULE OF FITS FOR BUSHING

Inside diameter   RC 3
Outside diameter   FN 2

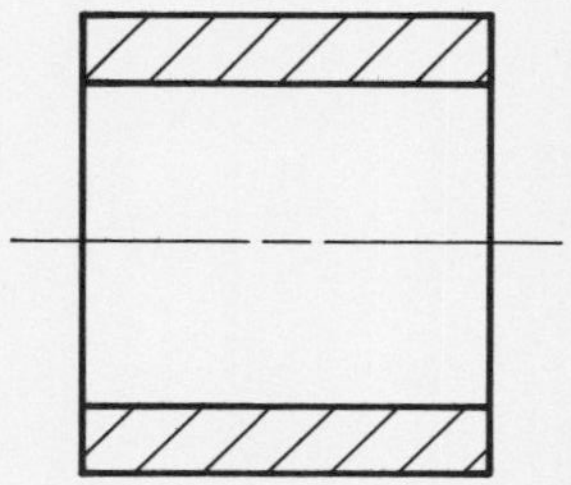

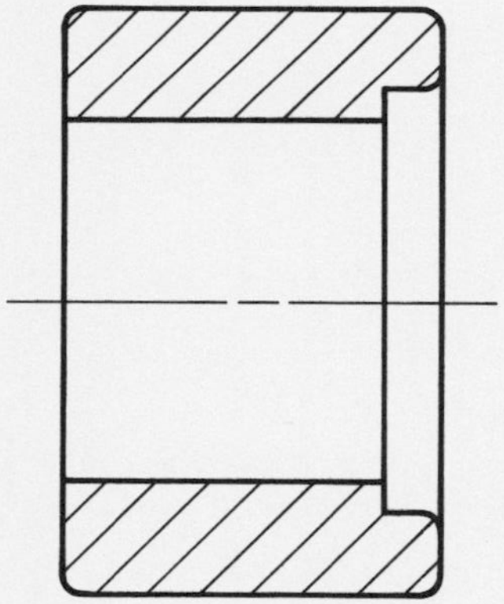

③ ROLLER
CRS – 1 REQD

② SPECIAL BOLT   CRS – 1 REQD

④ BUSHING
BRZ – 1 REQD

FILLETS AND ROUNDS 1.5 R
SCALE: 1 = 1

METRIC

| MATING PARTS DIMENSIONING | DRAWN BY | FILE NO. | DRAWING 67 |

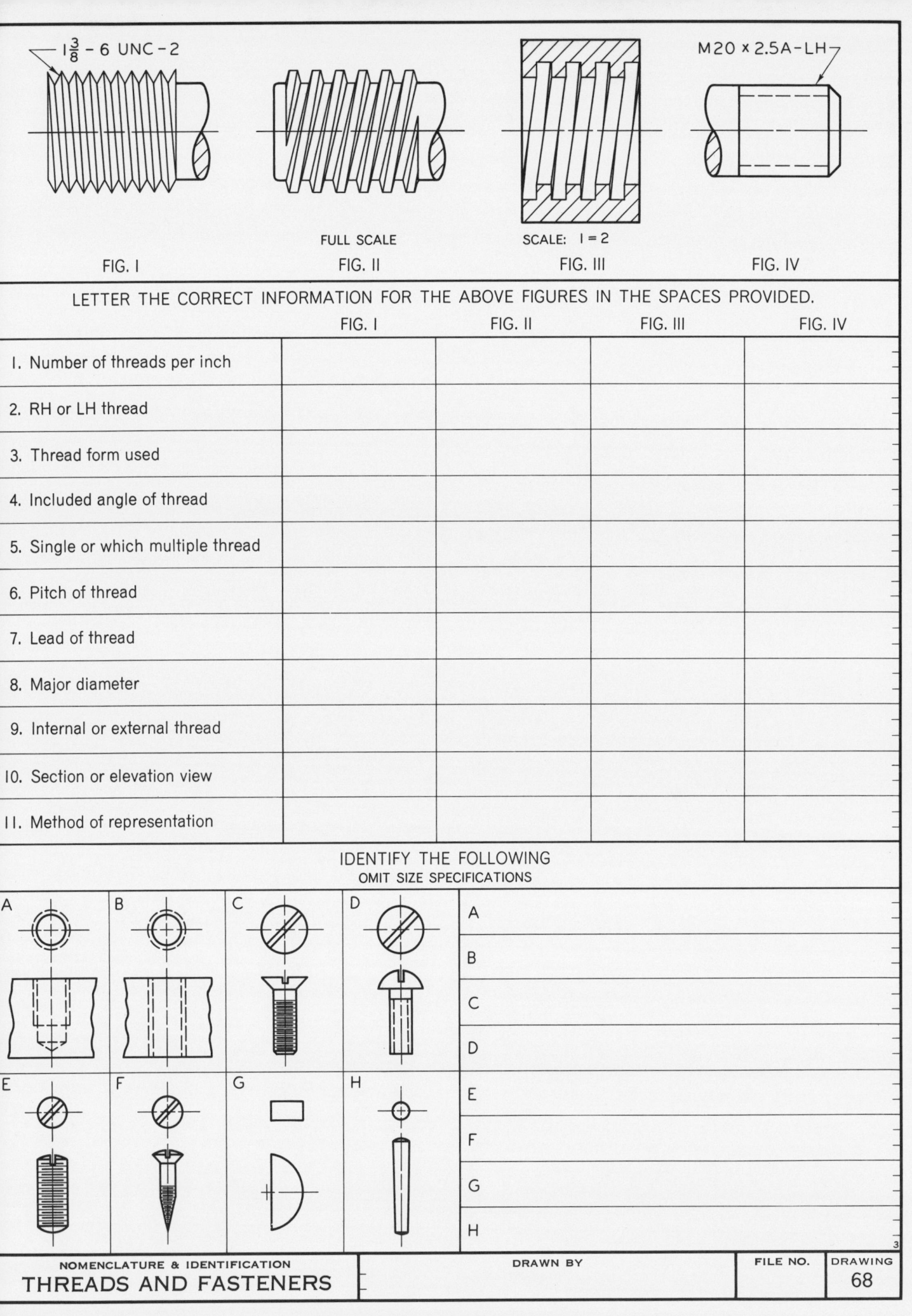

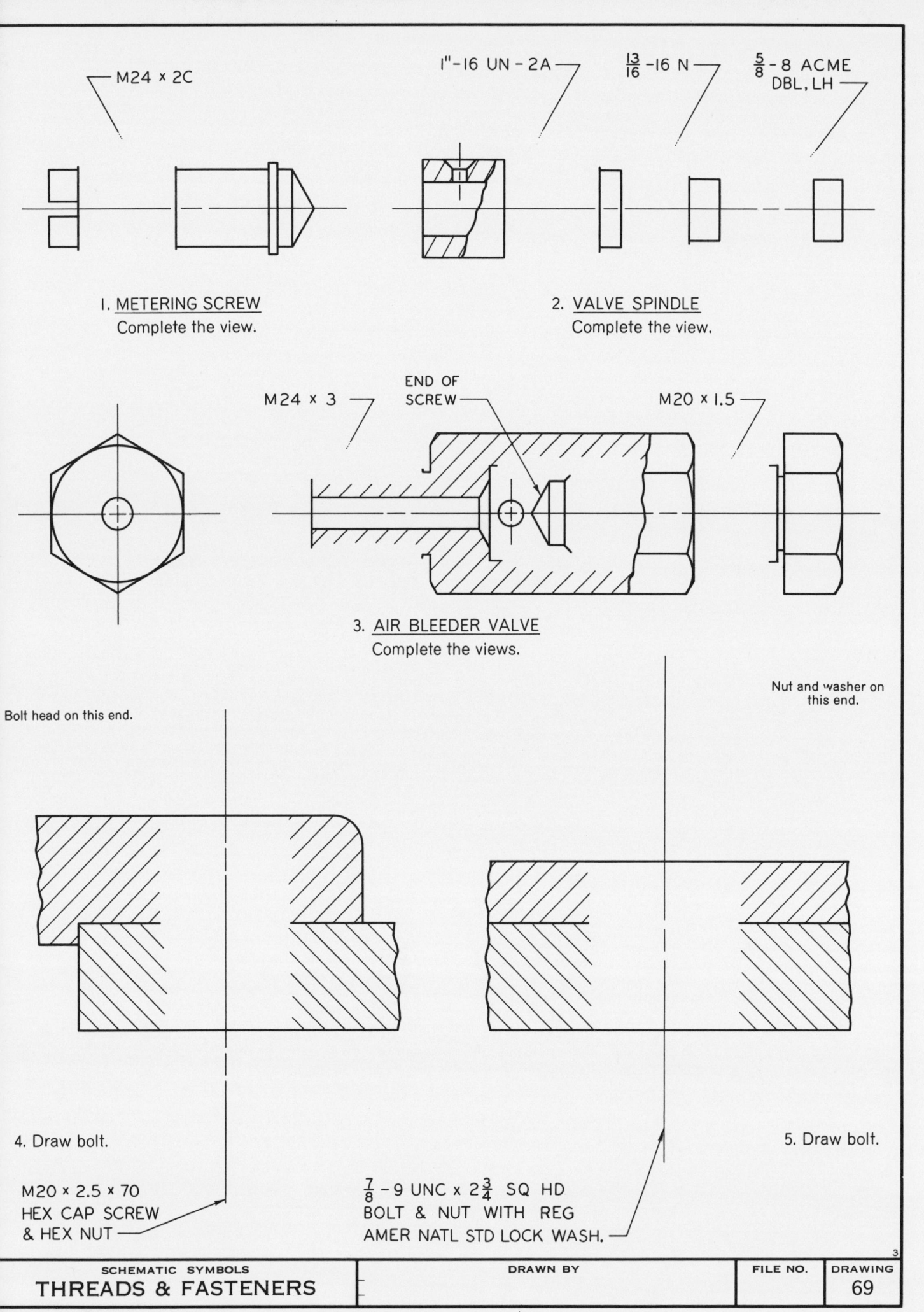

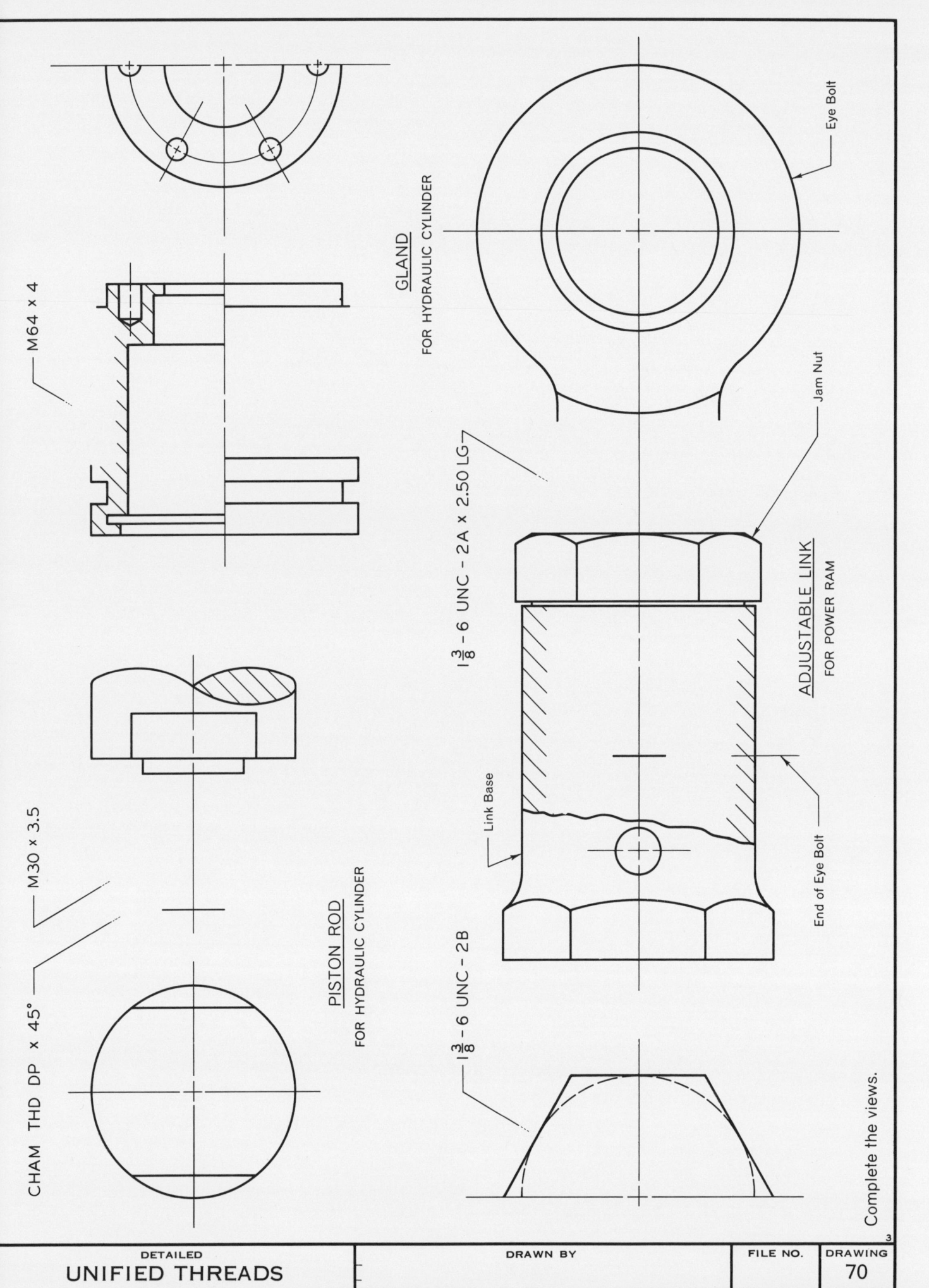

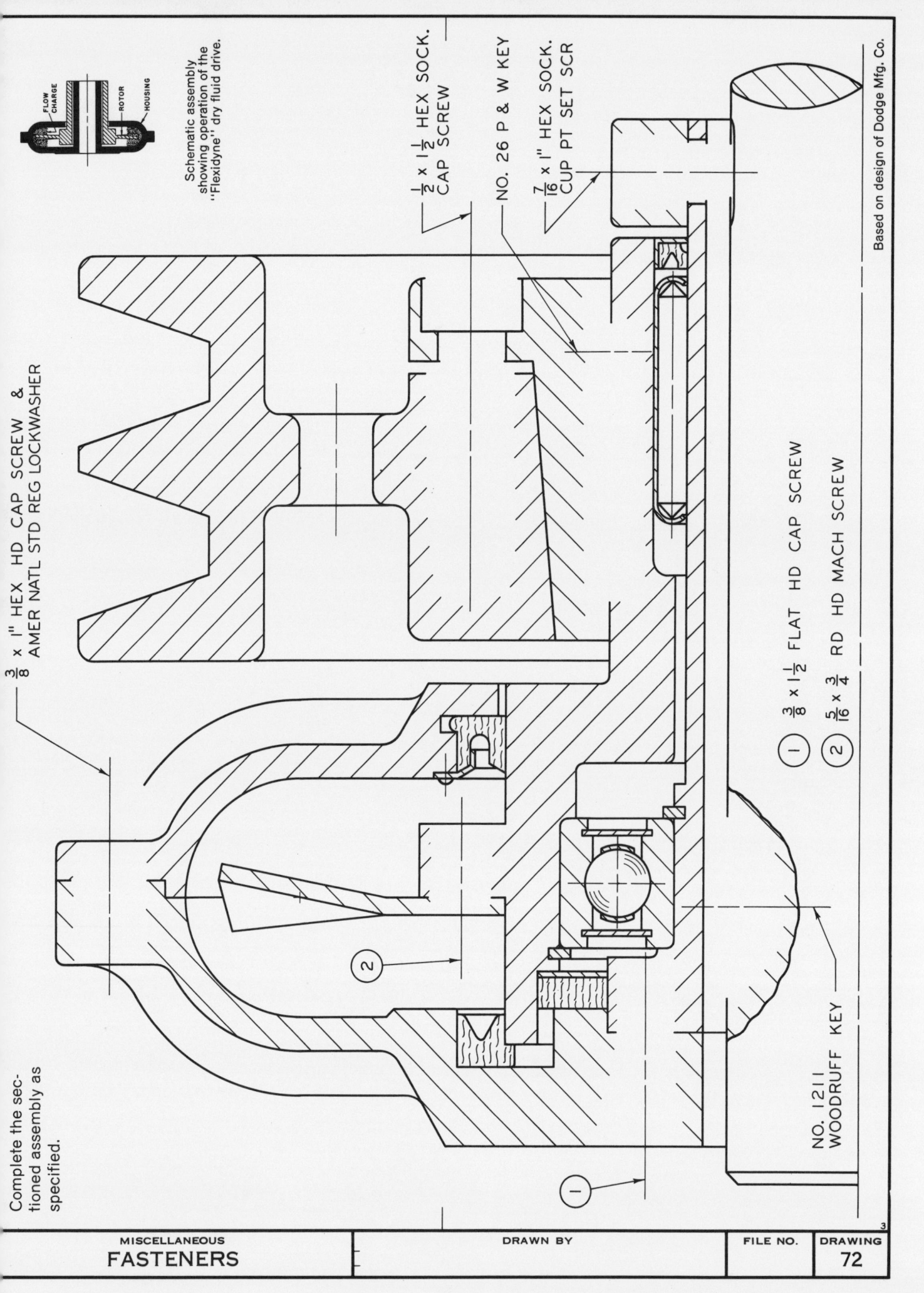

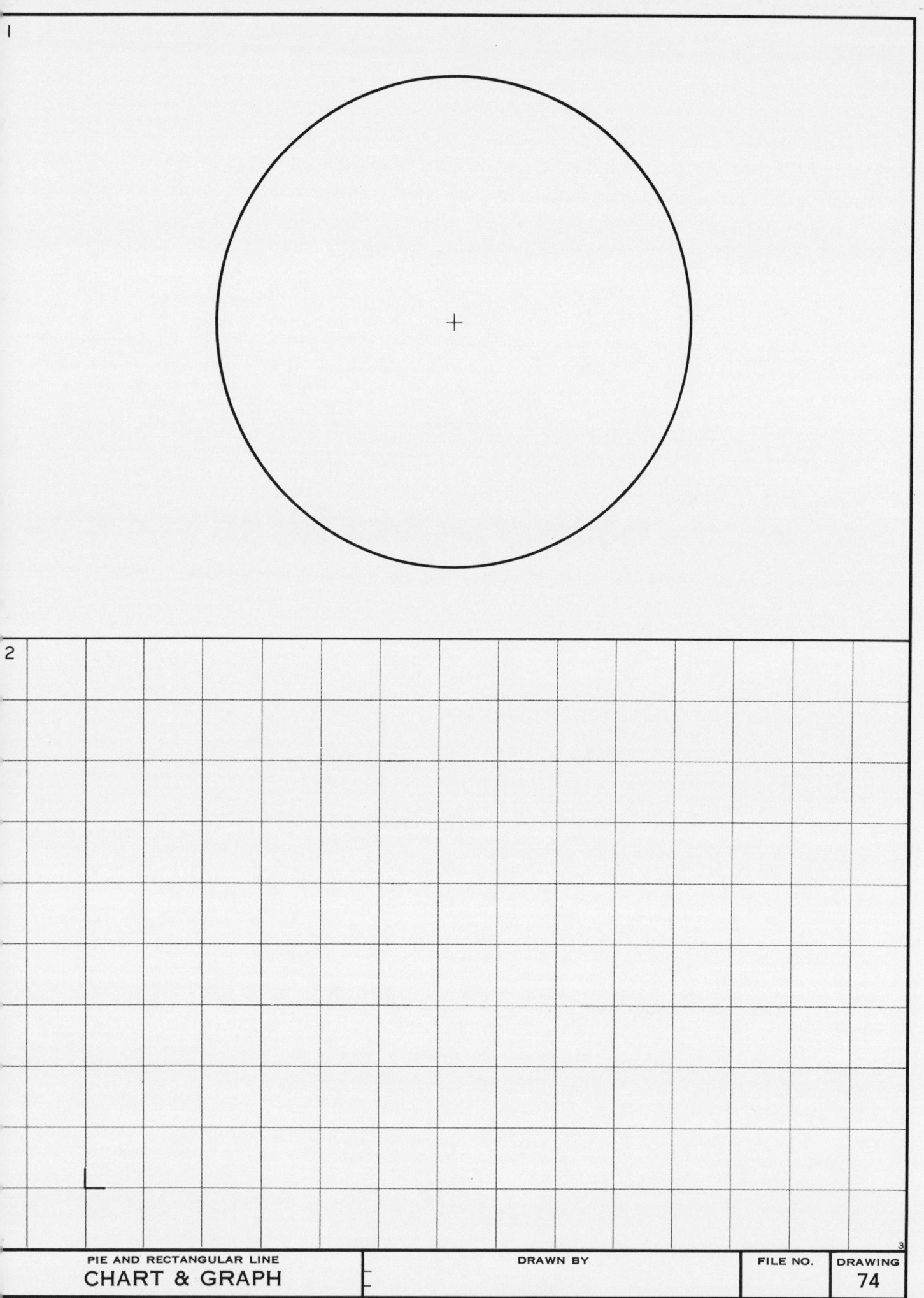

PIE AND RECTANGULAR LINE
CHART & GRAPH   DRAWN BY   FILE NO.   DRAWING
                                                     74

1

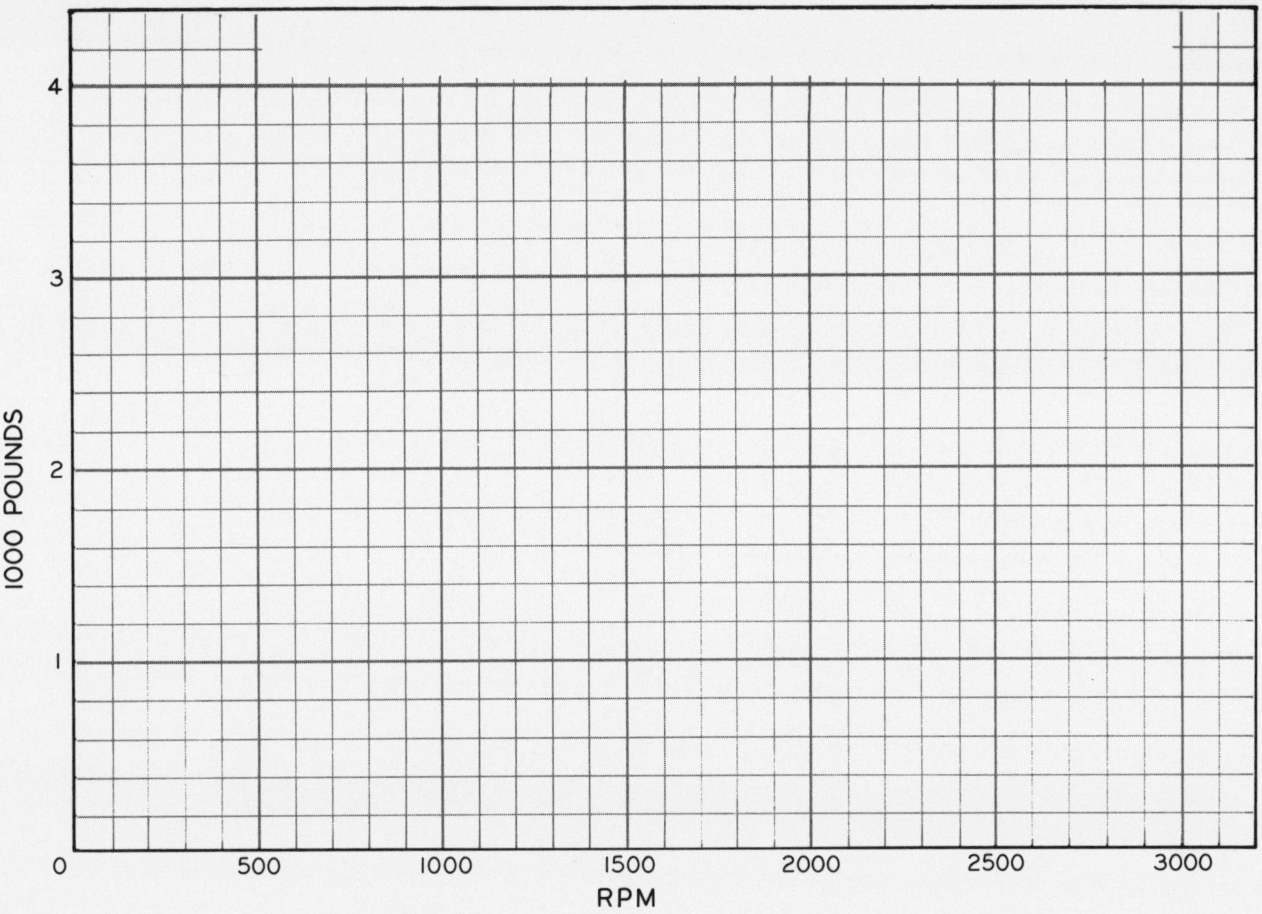

2

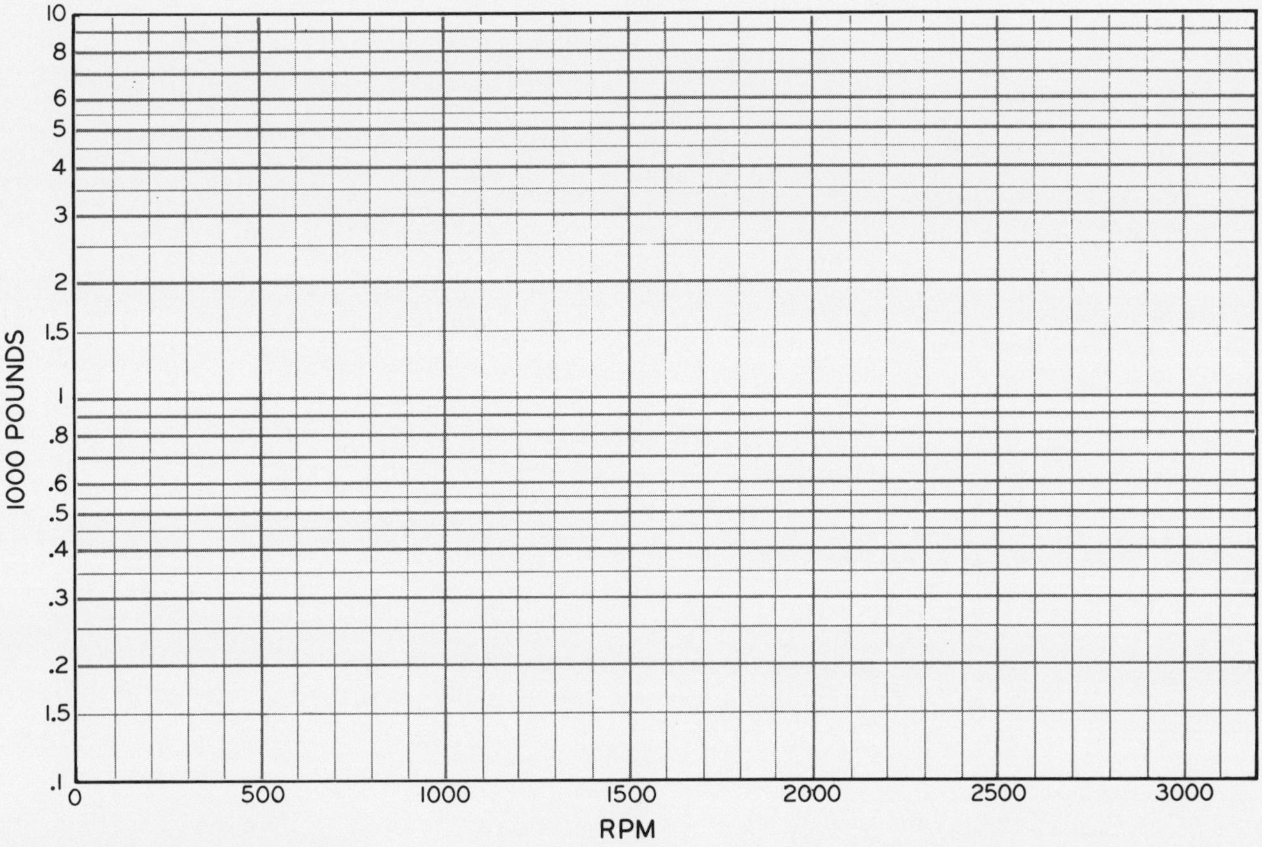

ARITHMETIC AND SEMI-LOGARITHMIC GRAPHS	DRAWN BY	FILE NO.	DRAWING 75

1. Find the values common to the equations:

   $y + \frac{x}{2} = 7$ & $2y - x = -8$.

2. Find the values common to the equations:

   $xy = 10$ & $2y - 3x = 8$.

GRAPHIC SOLUTIONS
SIMULTANEOUS EQUATIONS

DRAWN BY

FILE NO.

DRAWING 76

1. Find the roots of the equation:
$2x^2 + 2x - 7 = 0$

2. Find the roots of the equation:
$x^3 + 4x^2 - 5 = 0$

GRAPHIC SOLUTIONS
ROOTS OF EQUATIONS

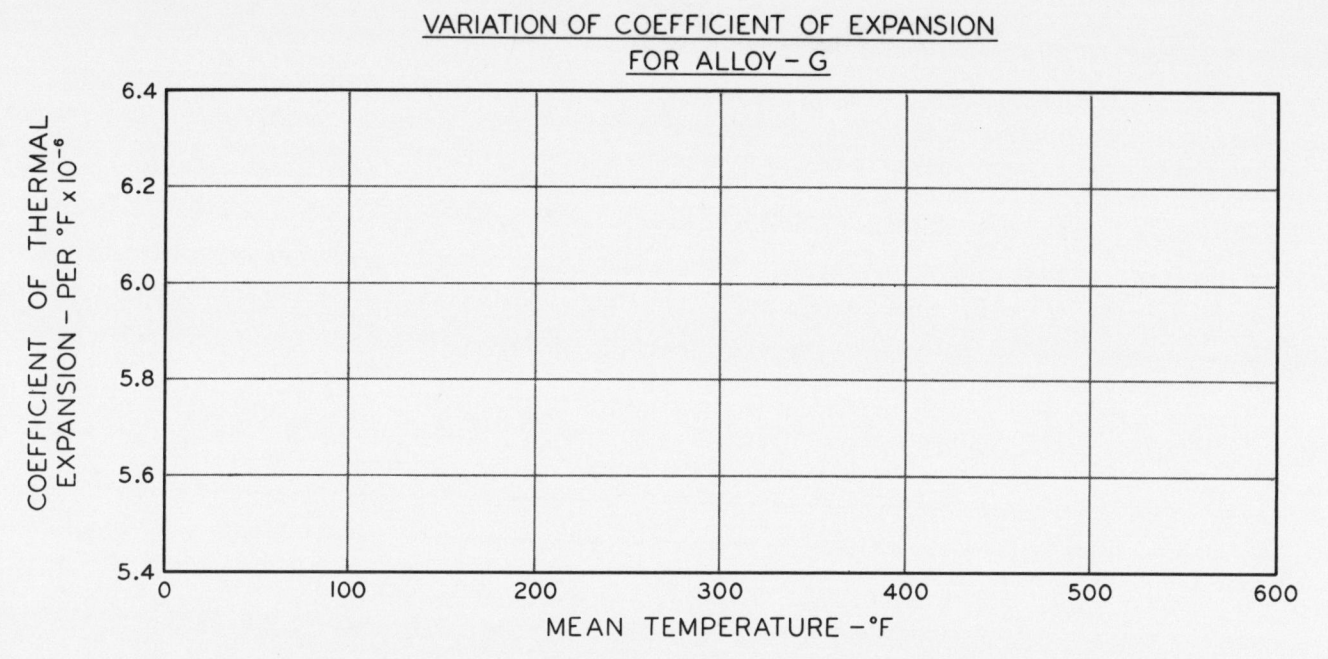

VARIATION OF COEFFICIENT OF EXPANSION FOR ALLOY - G

GRAPHIC SOLUTIONS
EMPIRICAL EQUATIONS

DRAWING 78

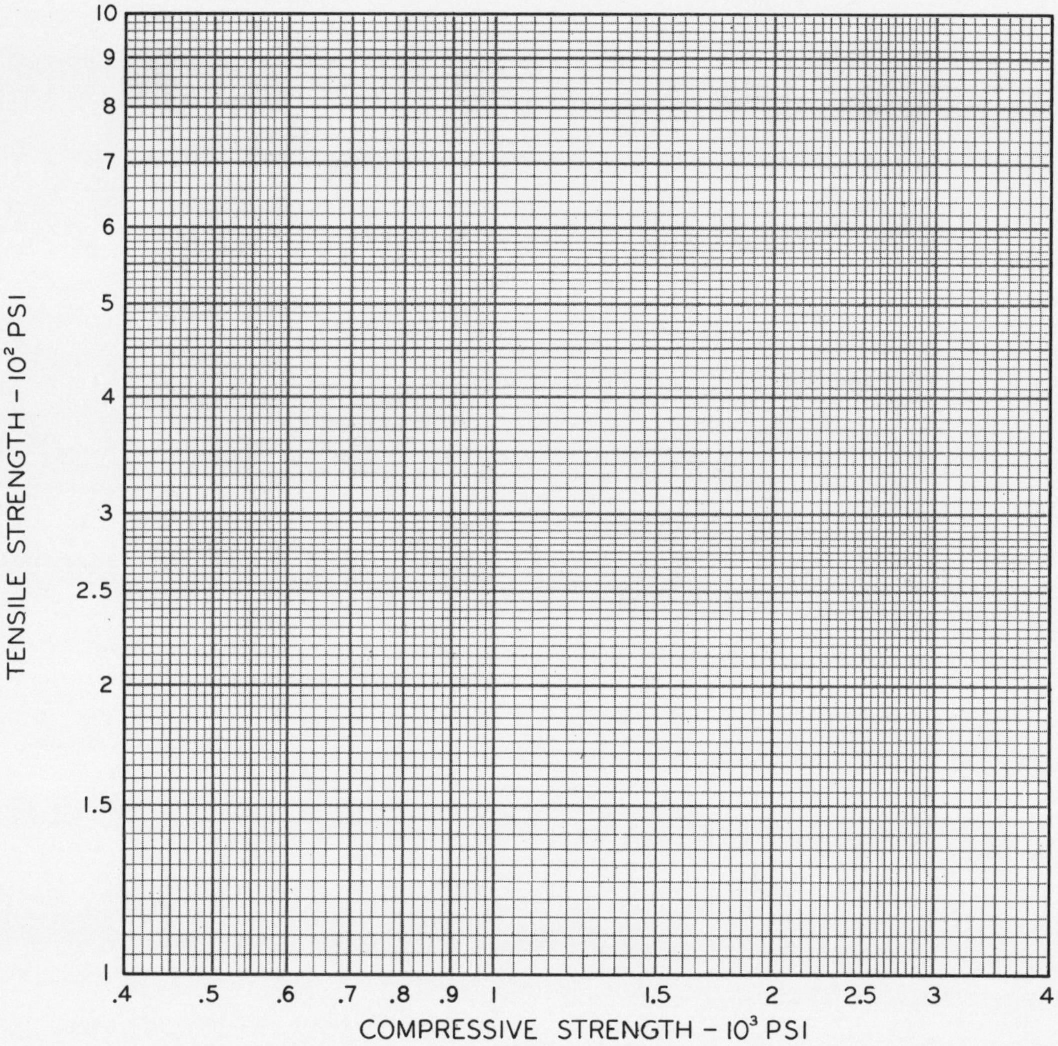

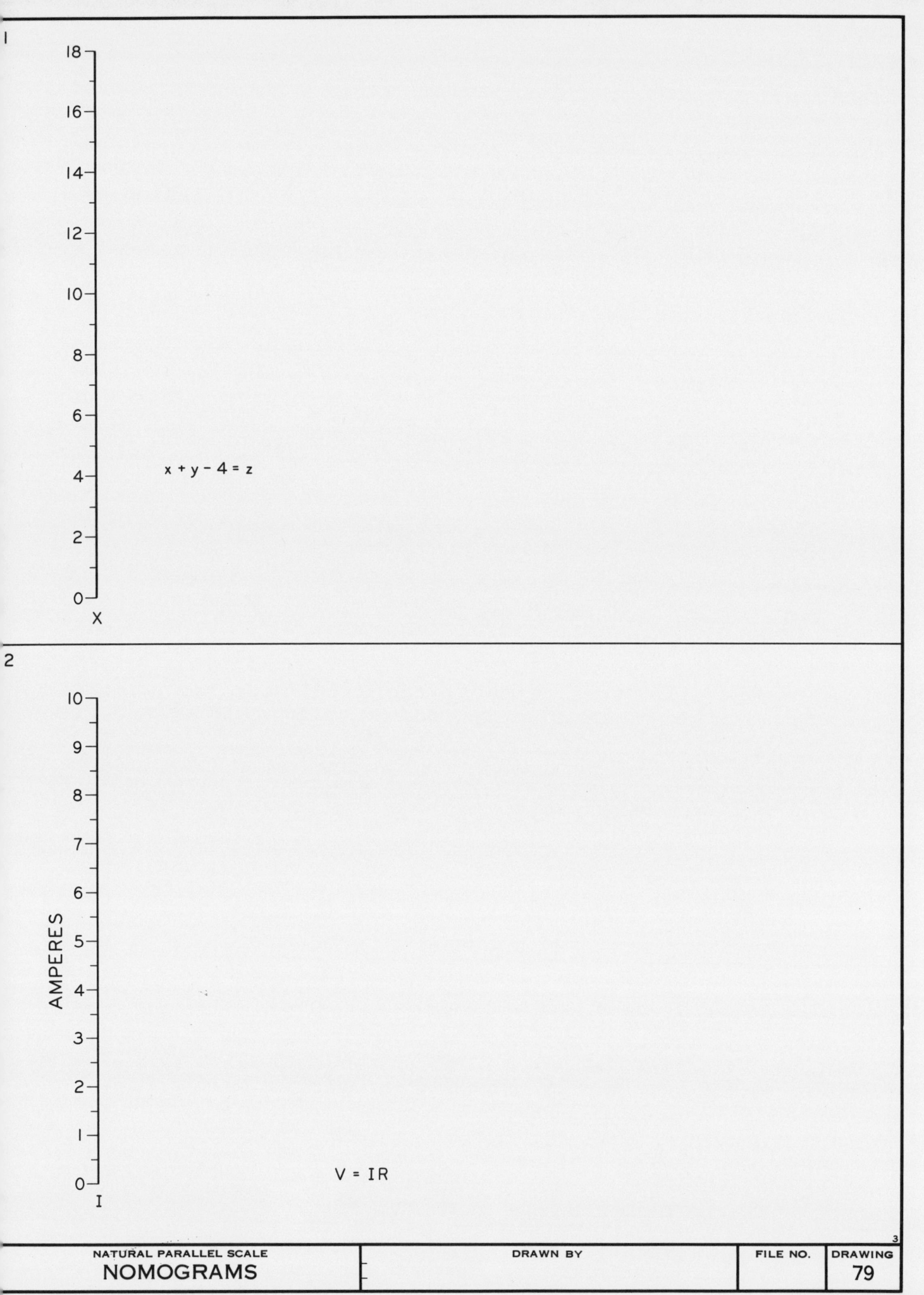

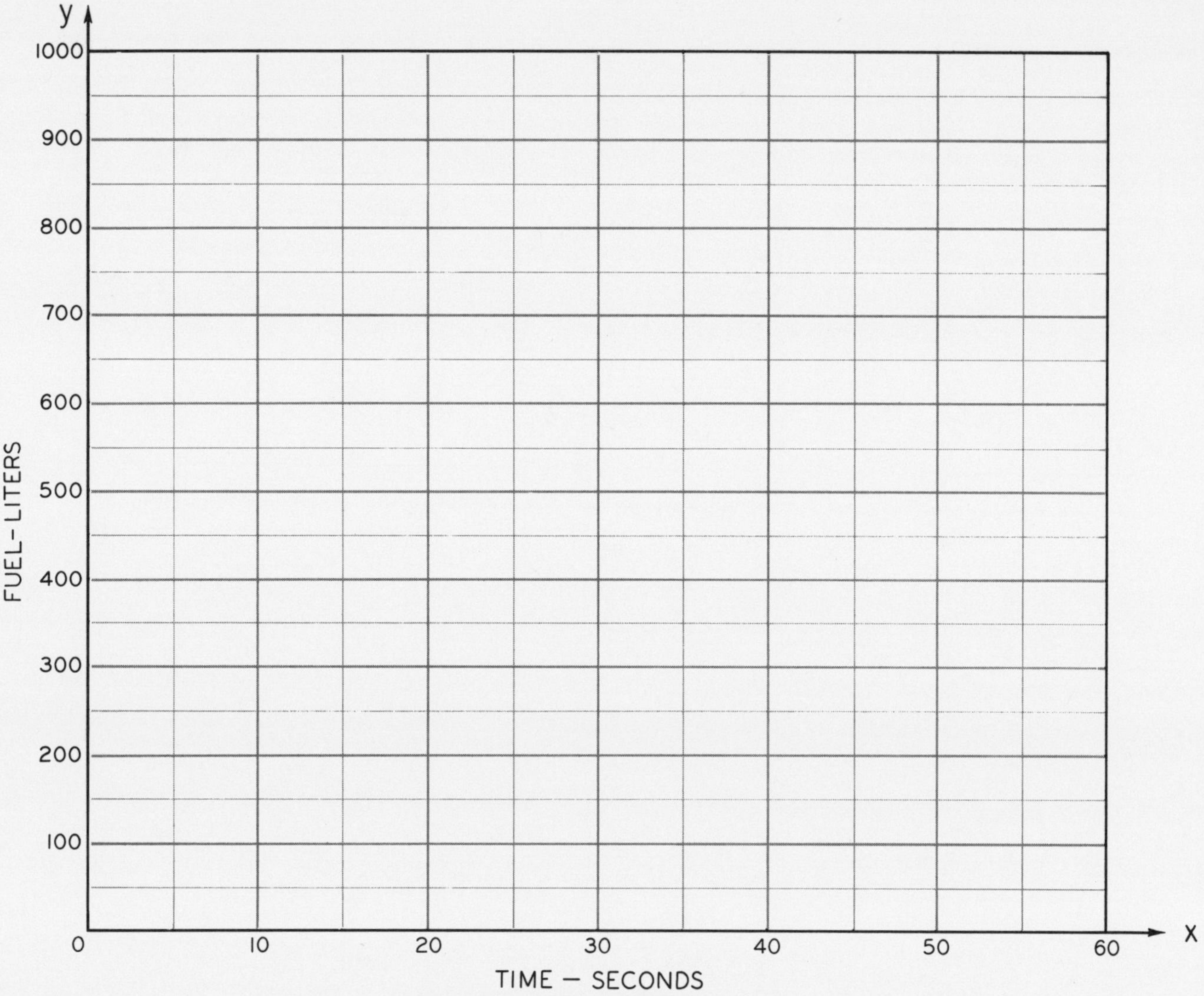

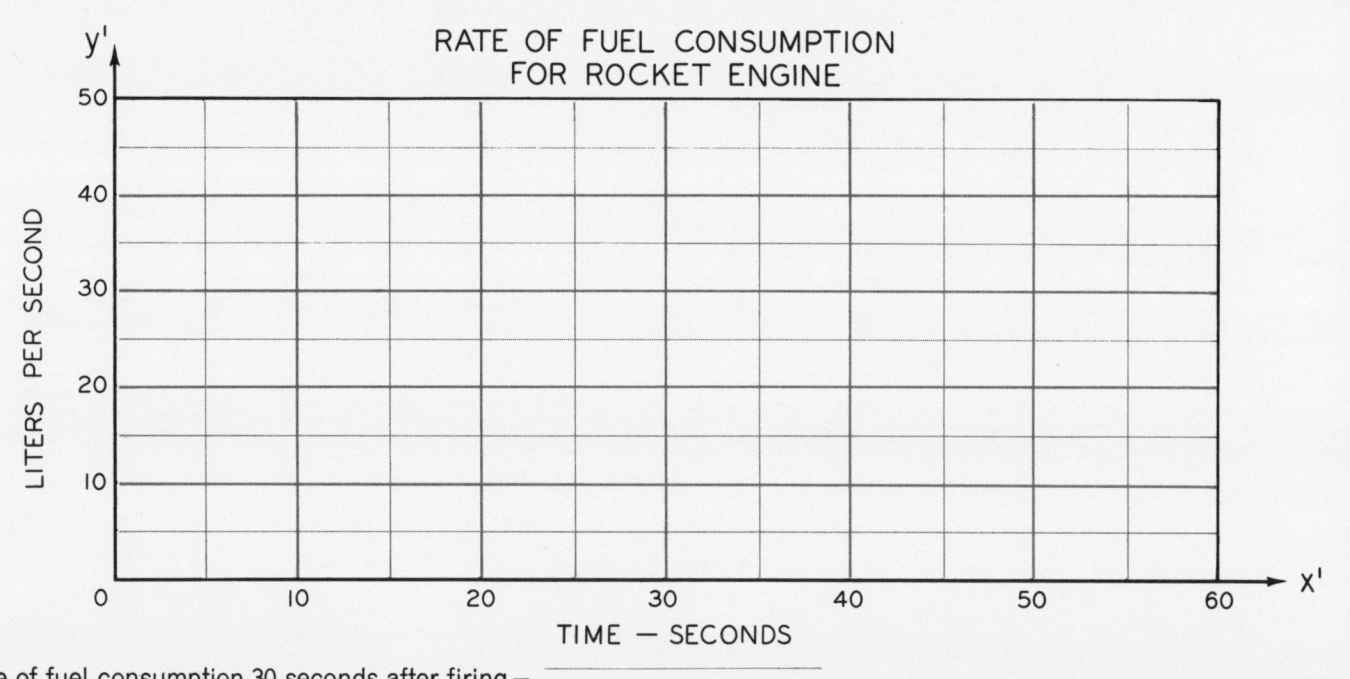

FUEL CONSUMPTION FOR
ROCKET ENGINE

RATE OF FUEL CONSUMPTION
FOR ROCKET ENGINE

Rate of fuel consumption 30 seconds after firing = ____________

3

DIFFERENTIATION GRAPHICAL CALCULUS	DRAWN BY	FILE NO.	DRAWING 80

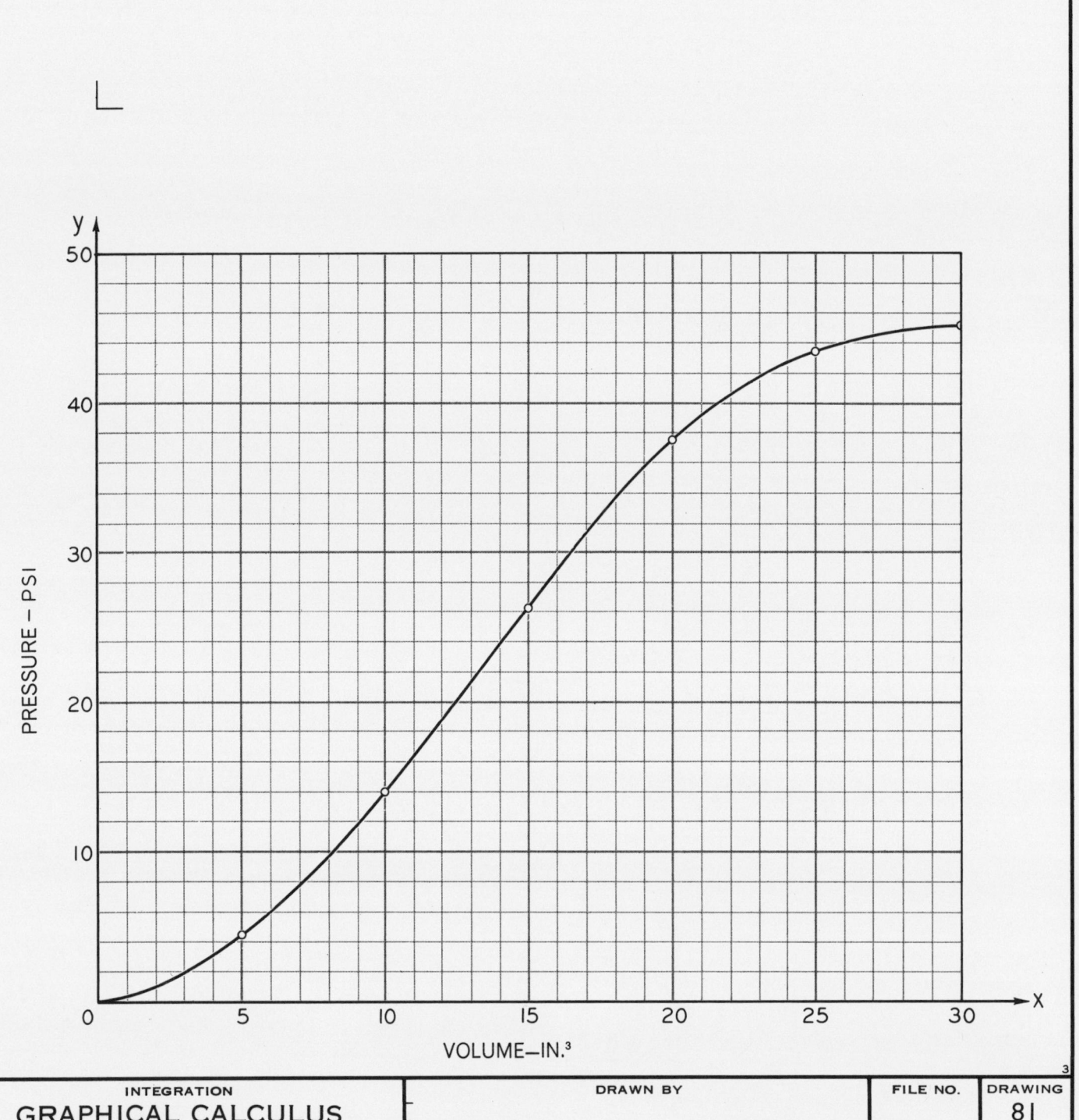

INTEGRATION	DRAWN BY	FILE NO.	DRAWING
GRAPHICAL CALCULUS			81

Complete the table of TERMS by entering the letter identifiers of the matching descriptions.

TERMS		Descriptions
CURSOR		A Handheld pointing device for pick and coordinate entry
DIGITIZER TABLET		B Computer program to perform specific tasks
GRAPHIC PRIMITIVE		C Counts in discrete steps or digits
PIXEL		D Smallest unit of digital information
RESOLUTION		E Collection of commands for selection
RASTER DISPLAY		F Device to convert analog picture to coordinate digital data
RAM		G Fundamental drawing entity
DEBUG		H Picture element dot in a display grid
HARD COPY		I Random Access Memory - volatile physical memory
ANALOG		J Continuous measurements without steps
DIGITAL		K Computer assisted engineering
CAE		L Group of 8 bits commonly used to represent a character
COMMAND		M Paper printout
PLOTTER		N Hand controlled lever used as input device
HARD DISK		O Smallest spacing between CRT display elements
VECTOR		P Convert an image into a proper display format
COORDINATE SYSTEM		Q Directed line segment with magnitude
MENU		R Flicker-free scanned CRT surface
BIT		S A bounded rectangular area on screen
MOUSE		T A visual tracking symbol
SOFTWARE		U Handheld photosensitive input device
JOY STICK		V Control signal
WINDOW		W Correct errors
TRANSFORM		X Non-volatile external storage device
BYTE		Y Hard copy device for vector drawing
LIGHTPEN		Z Common reference system for spatial relationships

COMPUTER-AIDED DRAFTING — TERMS AND DESCRIPTIONS

DRAWING 82

① Complete the table by defining X and Y coordinates of the given points.

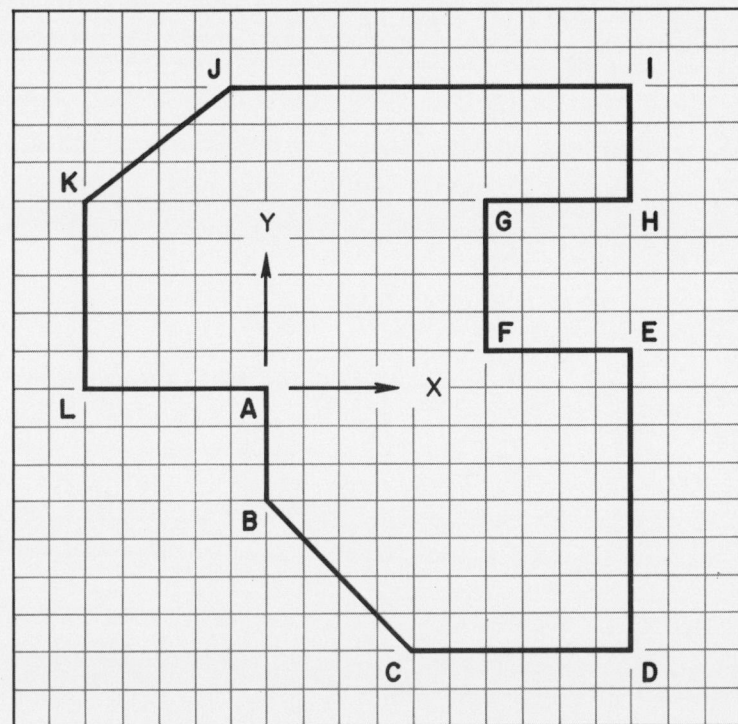

Point	Coordinate	
	X	Y
A	0	0
B	1	-3
C	4	-6
D	7	-6
E	7	-2
F	3	-2
G	3	2
H	7	2
I	7	6
J	-3	6
K	-5	4
L	-5	0

② Plot the given points on the grid and draw the view.

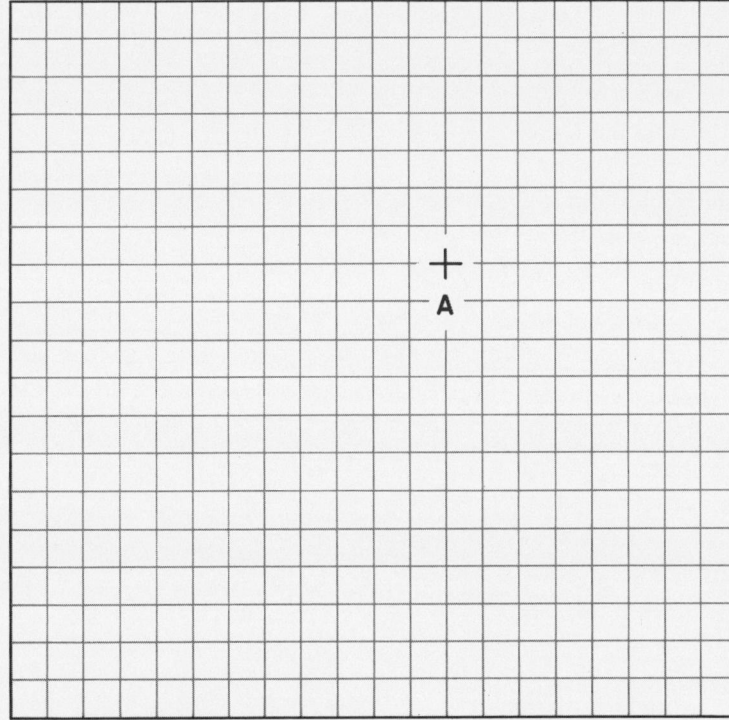

Point	Coordinate	
	X	Y
A	0	0
B	0	4
C	-5	4
D	-5	-2
E	-10	-6
F	-10	-10
G	-2	-10
H	6	-4
I	6	3
J	3	3
K	3	-2
L	0	-4

TWO-DIMENSIONAL COORDINATE PLOT
COMPUTER-AIDED DRAFTING

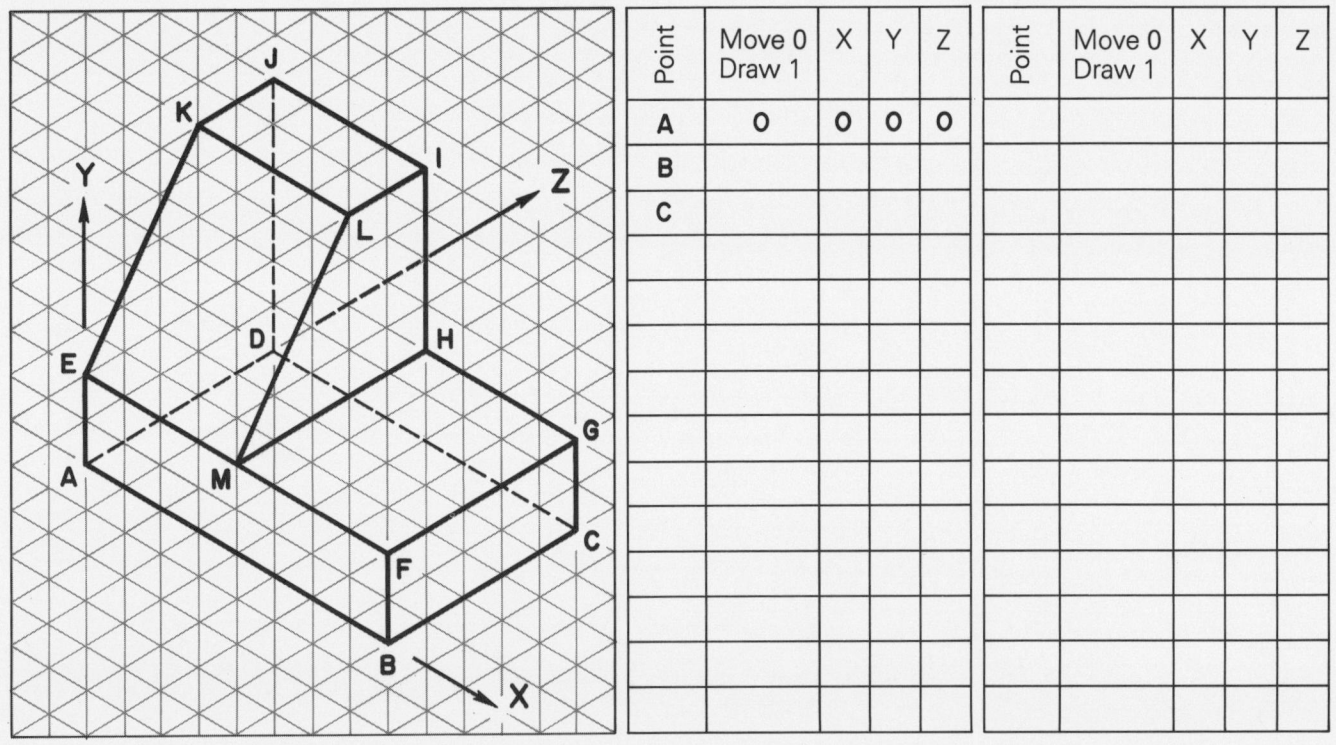

COMPUTER-AIDED DRAFTING

THREE-DIMENSIONAL COORDINATE PLOT

Drawing 84

① Complete the table for drawing the object.

② Draw the object based on the data given in the table.

Point	Move 0 Draw 1	X	Y	Z
A	0	0	0	0
B	1	0	5	0
C	1	0	5	5
D	1	0	0	5
E	1	6	0	5
F	1	6	0	0
A	1	0	0	0
D	1	0	0	5
G	0	-3	-4	0
L	1	-3	-1	0
K	1	-3	-1	5
J	1	-3	-4	5
I	1	0	-4	5
H	1	0	-4	0
G	1	-3	-4	0
J	1	-3	-4	5
L	0	-3	-1	0
B	1	0	5	0
K	0	-3	-1	5
C	1	0	5	5
H	0	0	-4	0
F	1	6	0	0
I	0	0	-4	5
E	1	6	0	5

Complete the table by entering the Menu Selections used for generating the drawing.

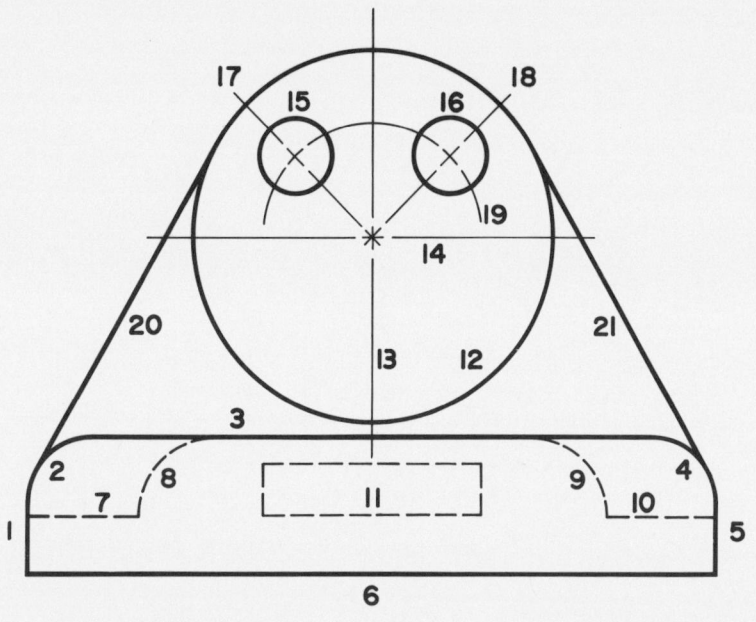

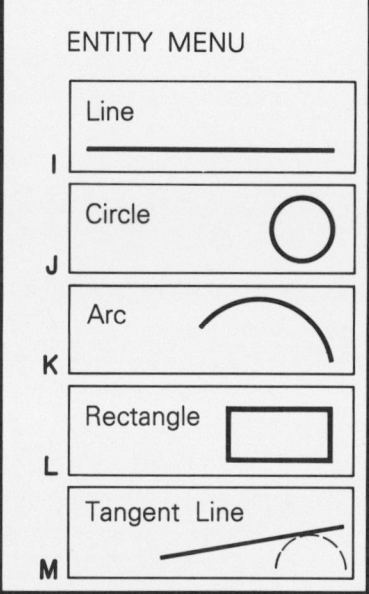

Entity	Line type menu selection	Entity menu selection	Construction menu selection
1			
2			
3			
4			
5			
6			
7			
8			
9			
10			
11			
12			
13			
14			
15			
16			
17			
18			
19			
20			
21			

MENU USAGE
COMPUTER-AIDED DRAFTING
DRAWN BY
FILE NO.
DRAWING 85

Description of VIEW COORDINATES

VIEW COORDINATES are the coordinate values of the object as assigned with respect to the computer screen, with X, Y and Z axes positioned as shown below. The coordinates remain the same irrespective of the view selected on the screen.

Axis	Position	Positive Direction
X	Horizontal	To the right
Y	Vertical	Toward the top
Z	Perpendicular to the screen	Outward from the screen

Description of WORLD COORDINATES

WORLD COORDINATES are the coordinate values of the object as assigned with respect to the axes of the object. The X, Y and Z axes are positioned as shown, such that for the top view the X axis is horizontal to the right, the Y axis is vertical to the top and the Z axis is perpendicular to the screen positioned outwards. The coordinates in relation to the screen change according to the view selected on the screen.

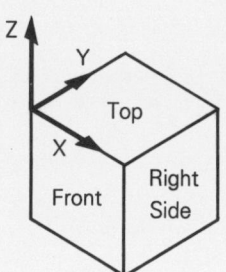

Complete the tables by entering the VIEW and WORLD COORDINATES of the given points of the object.

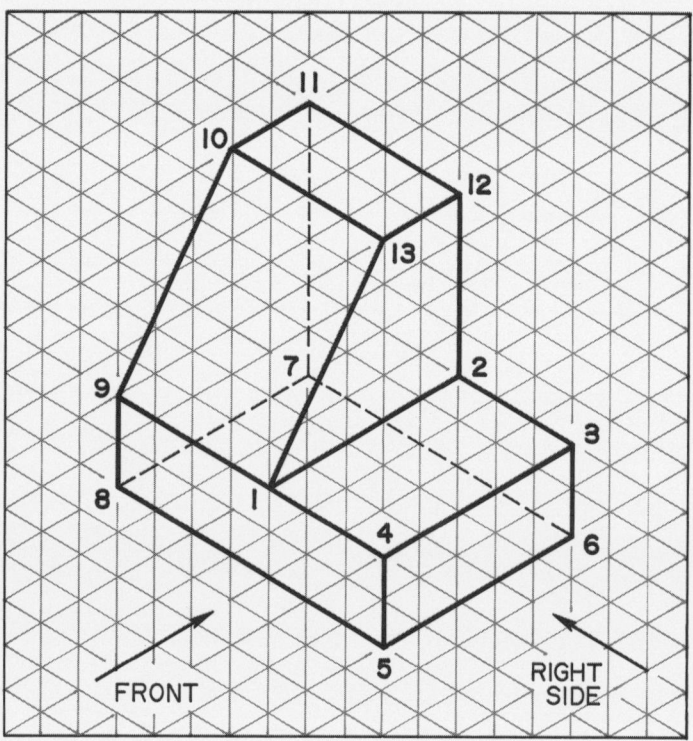

Points	FRONT VIEW					
	View Coordinates			World Coordinates		
	X	Y	Z	X	Y	Z
1						
2						
3						
4						
5						
6						
7						
8						
9						
10						
11						
12						
13						

Points	RIGHT SIDE VIEW					
	View Coordinates			World Coordinates		
	X	Y	Z	X	Y	Z
1						
2						
3						
4						
5						
6						
7						
8						
9						
10						
11						
12						
13						

COORDINATE SYSTEMS
COMPUTER-AIDED DRAFTING

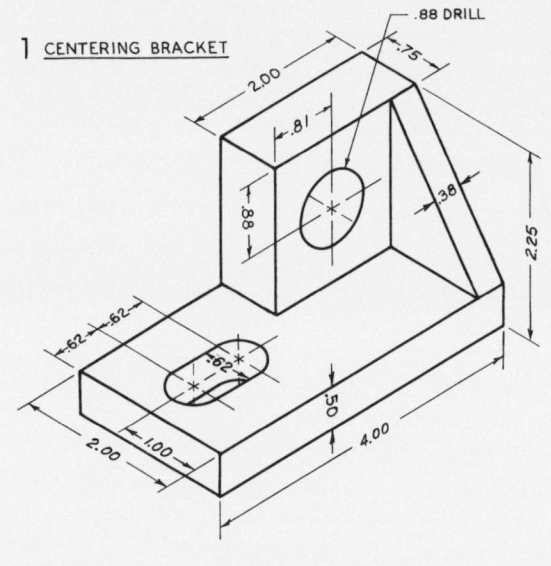
1. CENTERING BRACKET

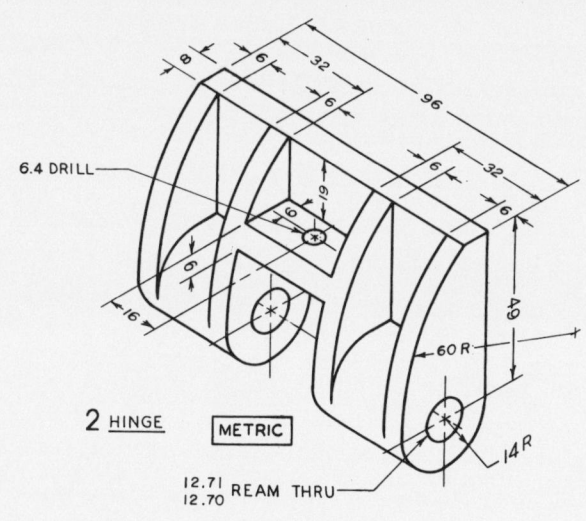
2. HINGE — METRIC

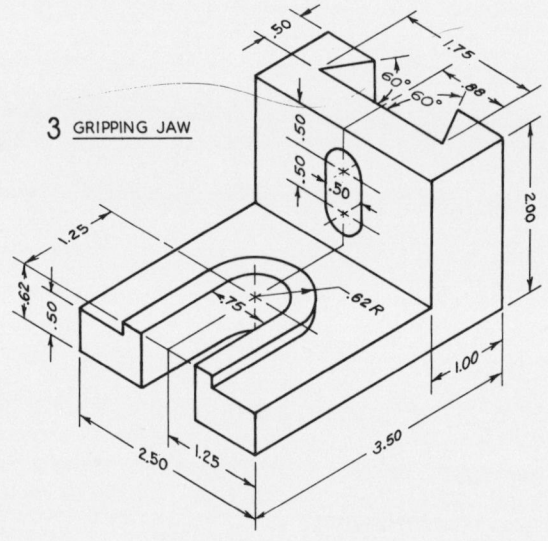
3. GRIPPING JAW

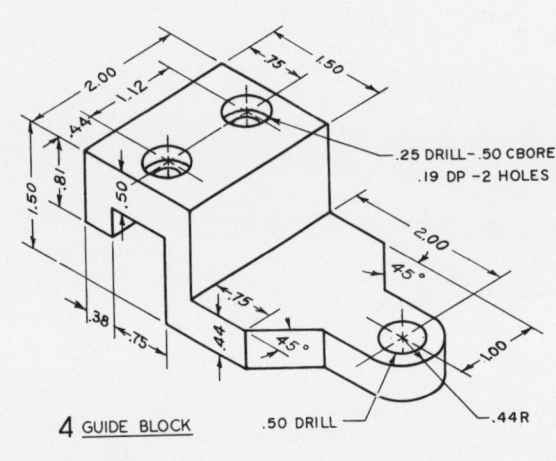
4. GUIDE BLOCK

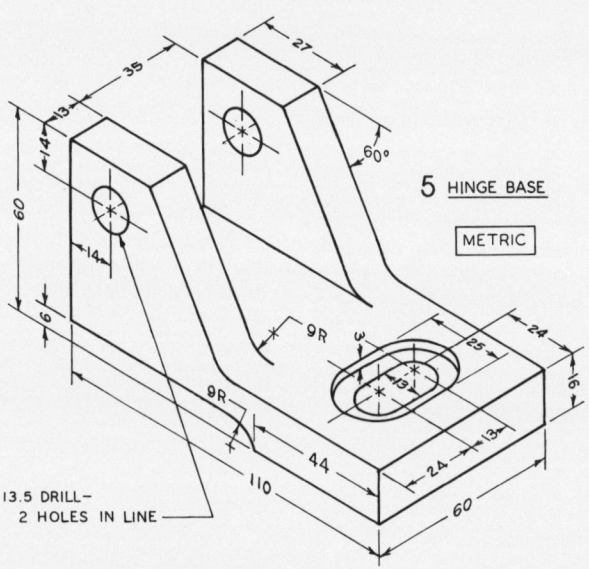
5. HINGE BASE — METRIC

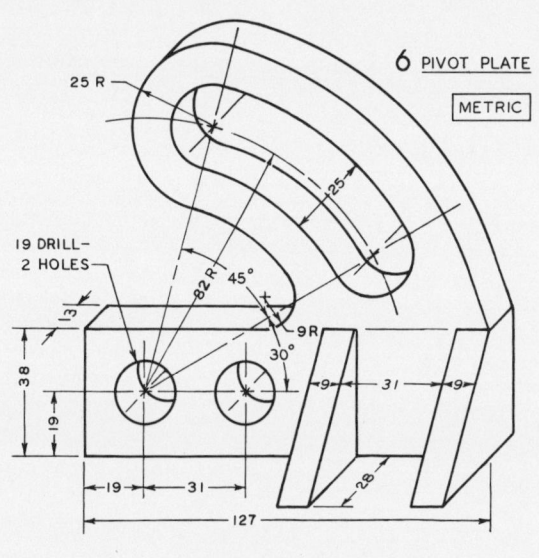
6. PIVOT PLATE — METRIC

DETAIL DRAWINGS — DRAW OR SKETCH THE NECESSARY VIEWS OF THE OBJECT ASSIGNED. DIMENSION COMPLETELY

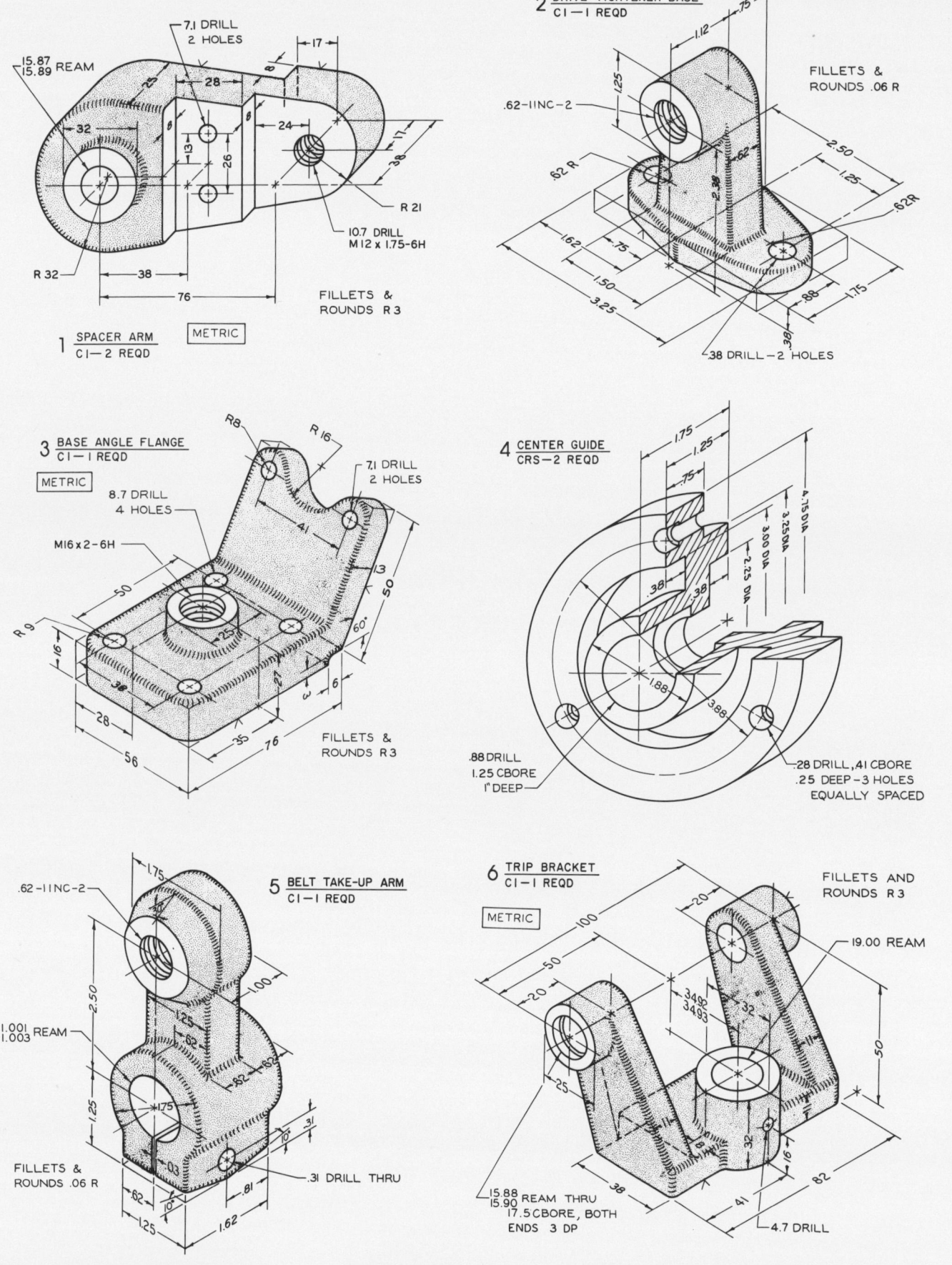

DETAIL DRAWINGS

DRAW OR SKETCH THE NECESSARY VIEWS OF THE OBJECT ASSIGNED. DIMENSION COMPLETELY

DRAWING 88

| | DRAWN BY | FILE NO. | DRAWING |

	DRAWN BY	FILE NO.	DRAWING

| | | DRAWN BY | | FILE NO. | DRAWING |

| | DRAWN BY | FILE NO. | DRAWING |

| | DRAWN BY | FILE NO. | DRAWING |

| | DRAWN BY | FILE NO. | DRAWING |

DRAWN BY | FILE NO. | DRAWING

| | DRAWN BY | FILE NO. | DRAWING |

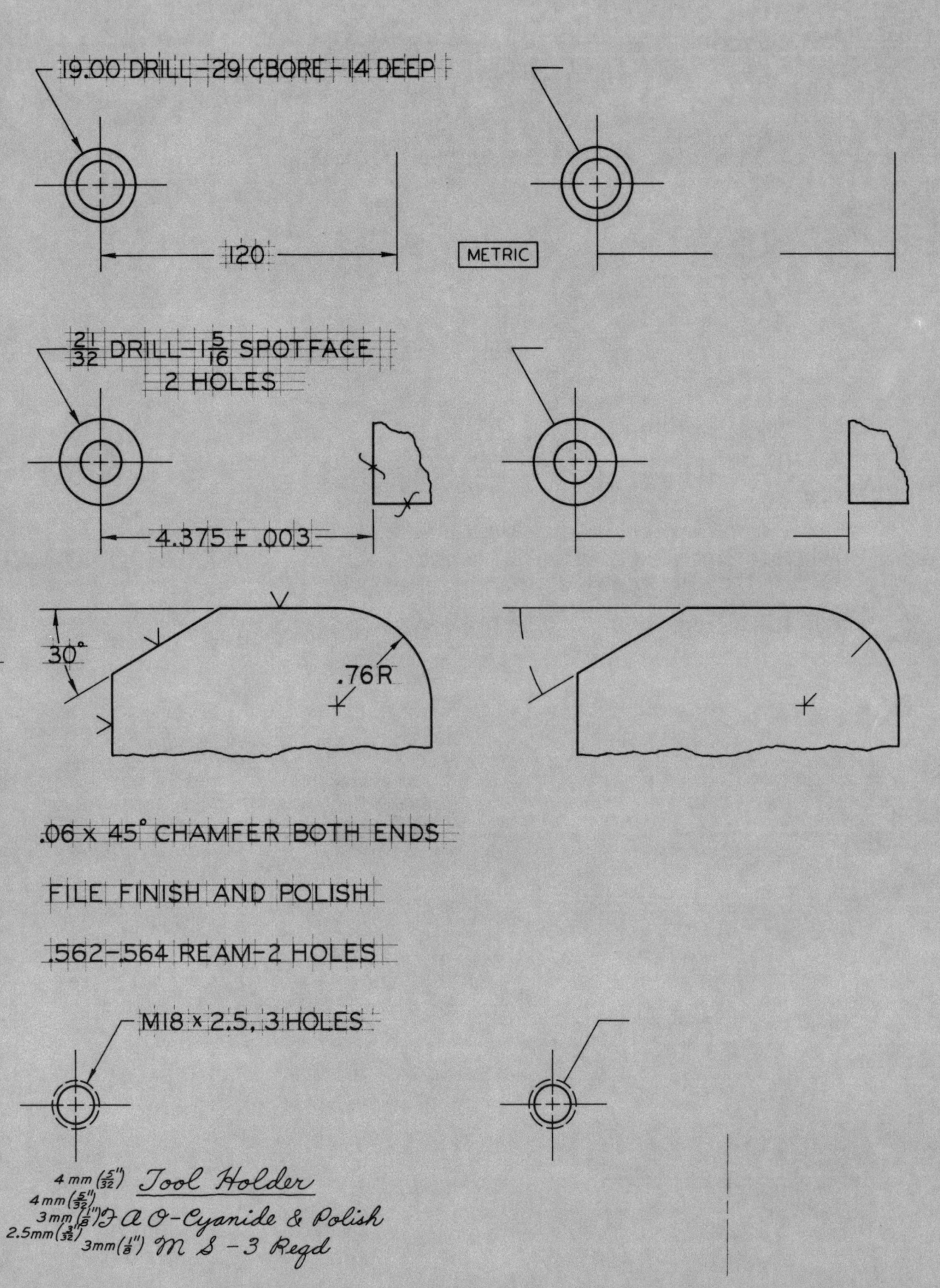

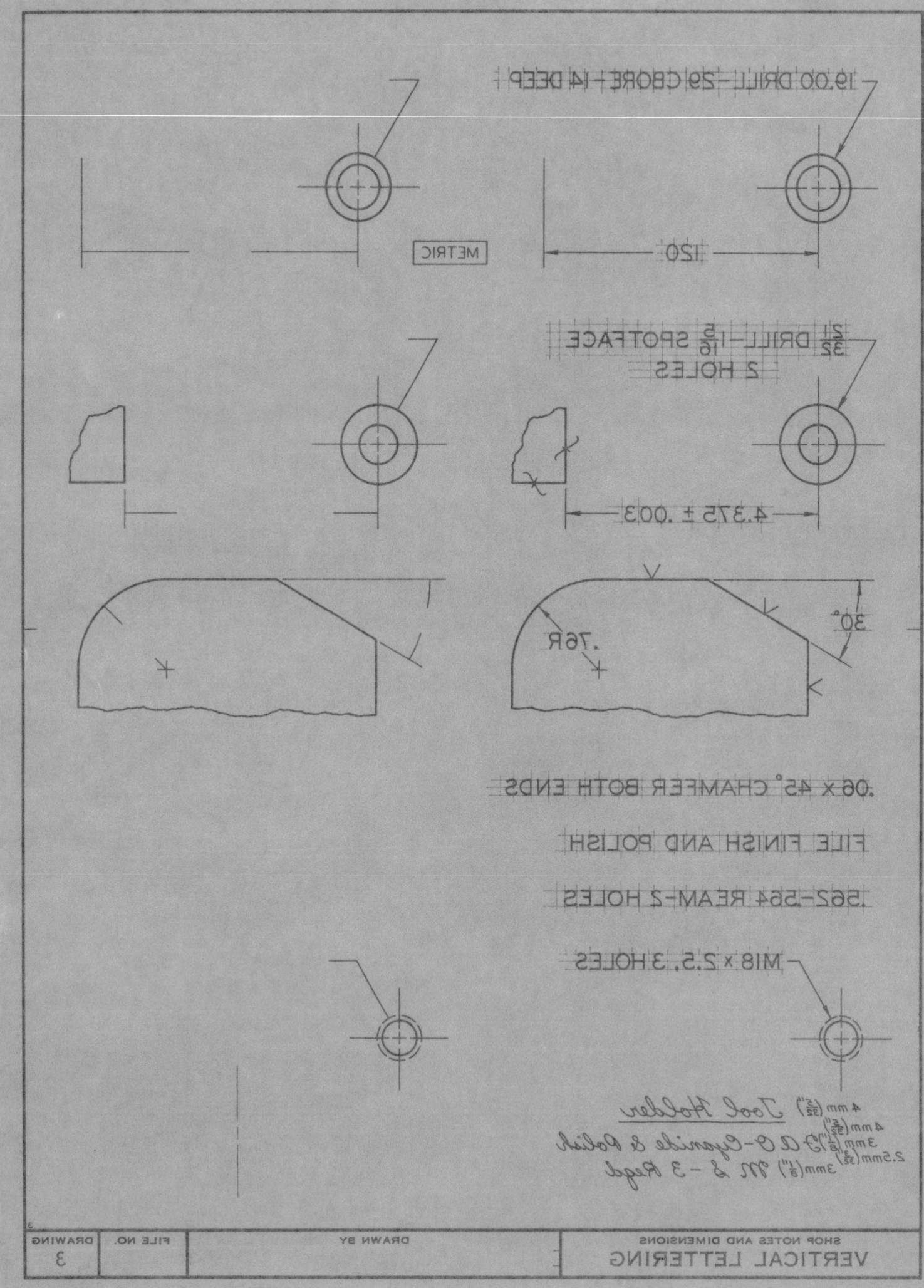

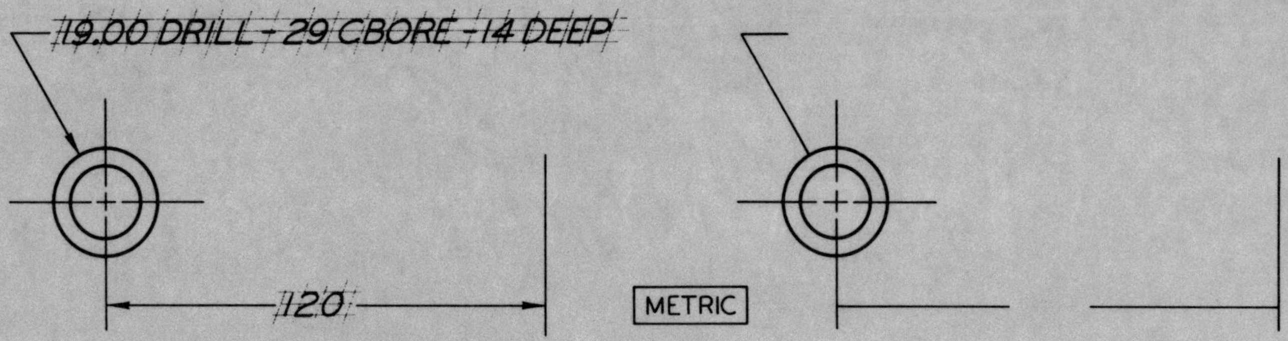

19.00 DRILL - 29 CBORE - 14 DEEP

|←——— 120 ———→|

METRIC

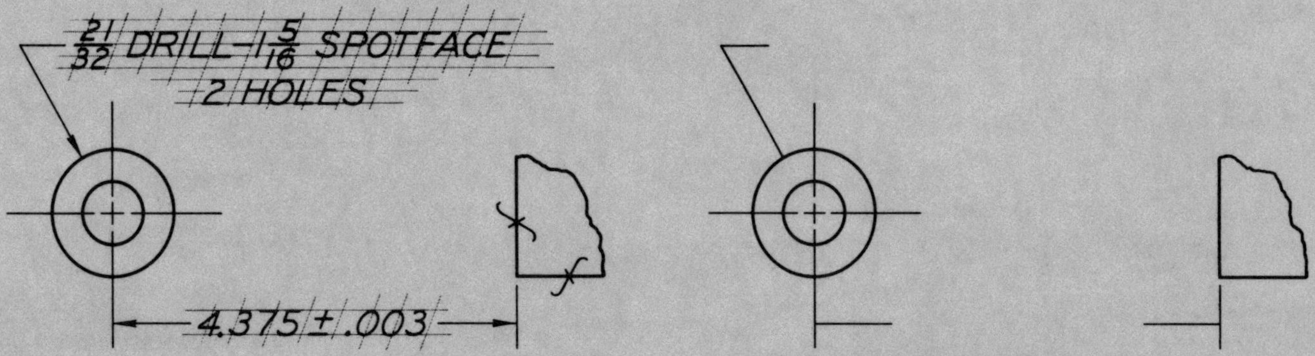

21/32 DRILL - 1 5/16 SPOTFACE
2 HOLES

|←——— 4.375 ±.003 ———→|

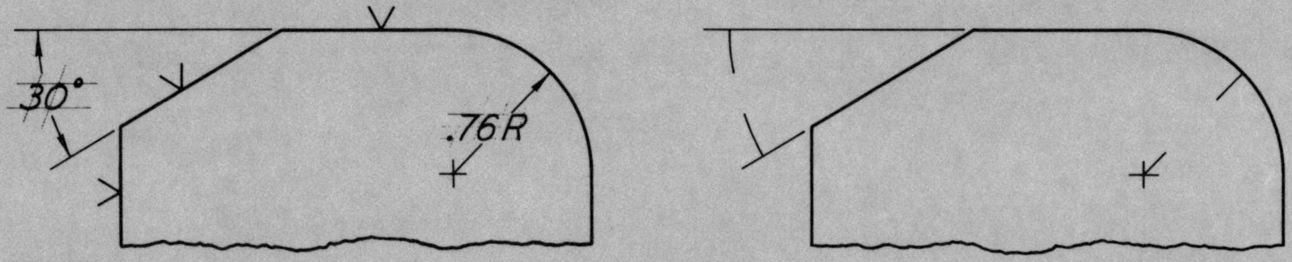

30°   .76R

.06 × 45° CHAMFER BOTH ENDS

FILE FINISH AND POLISH

.562-.564 REAM - 2 HOLES

M18 × 2.5, 3 HOLES

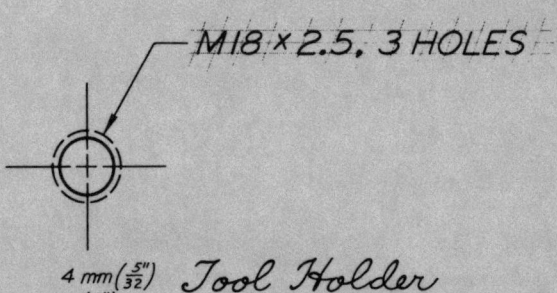
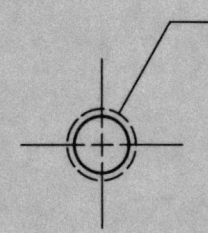

4 mm (5/32")   *Tool Holder*
4 mm (5/32")
3 mm (1/8")   *F A O - Cyanide & Polish*
2.5 mm (3/32")
3 mm (1/8")   *M S - 3 Reqd*

| SHOP NOTES AND DIMENSIONS / INCLINED LETTERING | DRAWN BY | FILE NO. | DRAWING 7 |

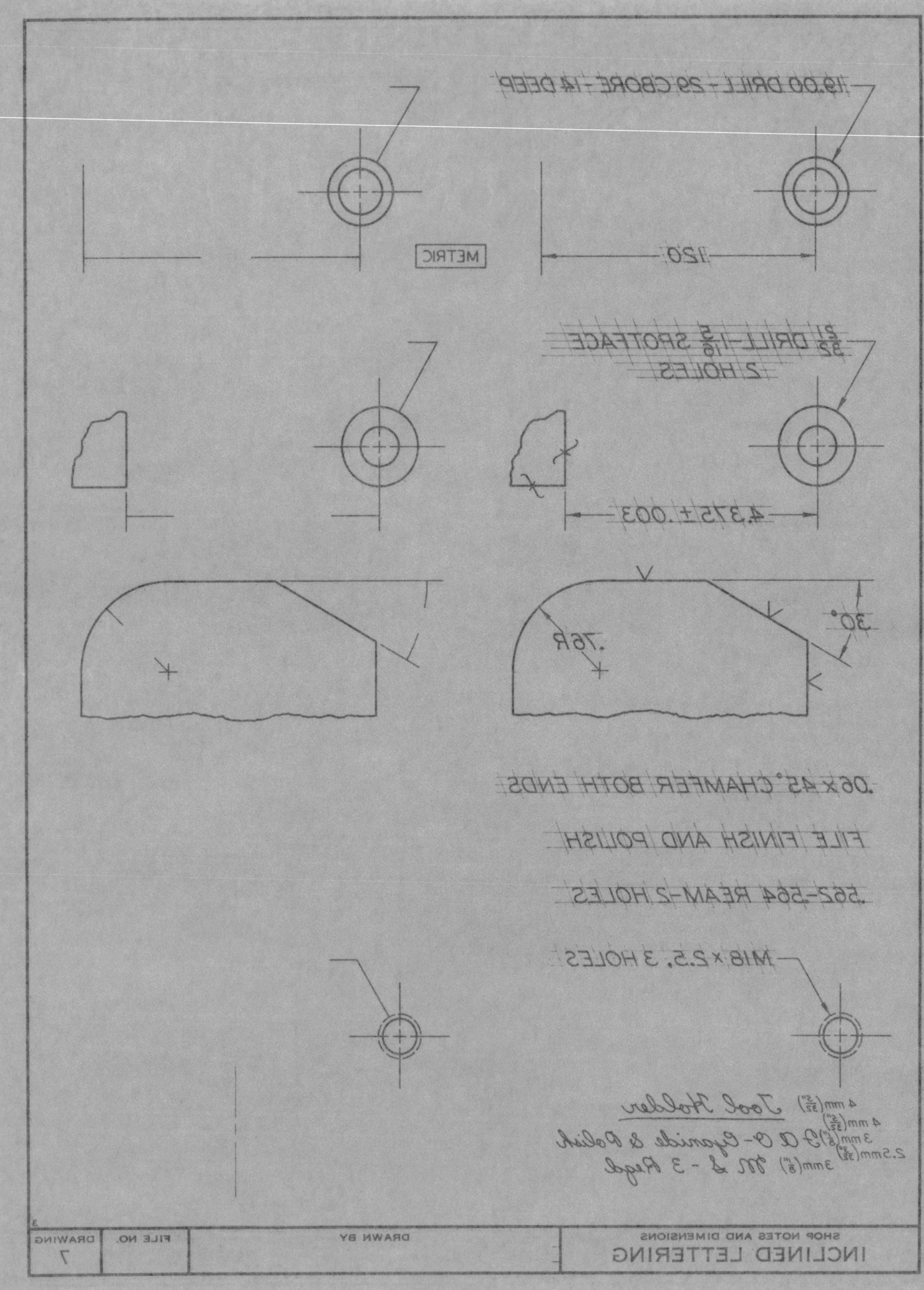

1

A — SCALE: 12"= 1'-0   L = _____

B — SCALE: 6"= 1'-0   L = 11 9/16"

C — SCALE: 1"= 10'   L = _____

D — SCALE: 1"= 20'   L = 97.5'

E — SCALE: 1"= 600'   L = 3100.0'

F — SCALE: 1mm = 1mm   L = _____

G — SCALE: 1mm = 10mm   L = 1315.0mm

H — SCALE: 1mm = 2mm   L = 286.0mm

J — SCALE: _____   L = _____

K — SCALE: _____   L = _____

L — SCALE: _____   L = _____

Measure or draw the above lengths as indicated and record answers in appropriate spaces.

2

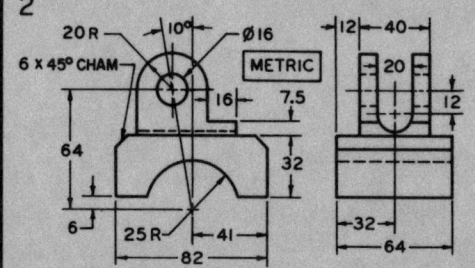

ANCHOR BRACKET

CRS—6 REQD
SCALE: 1 = 1

Draw the two views.
Omit dimensions.

SCALES AND LAYOUT
MECHANICAL DRAWING

DRAWN BY

FILE NO.

DRAWING 10

MECHANICAL DRAWING
SCALES AND LAYOUT

	A	SCALE : 12"= 1'–0	L = _____
	B	SCALE : 6" = 1'–0	L = 11 9/16"
	C	SCALE : 1" = 10'	L = _____
	D	SCALE : 1" = 20'	L = 97.5'
	E	SCALE : 1" = 600'	L = 3100.0'
	F	SCALE : 1mm = 1mm	L = _____
	G	SCALE : 1mm = 10mm	L = 1315.0mm
	H	SCALE : 1mm = 2mm	L = 286.0mm
	J	SCALE : _____	L = _____
	K	SCALE : _____	L = _____
	L	SCALE : _____	L = _____

Measure or draw the above lengths as indicated and record answers in appropriate spaces.

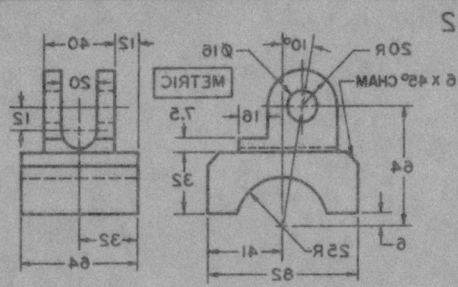

ANCHOR BRACKET
CRS—6 REQD
SCALE: 1 = 1

Draw the two views.
Omit dimensions.

DRAWING 10

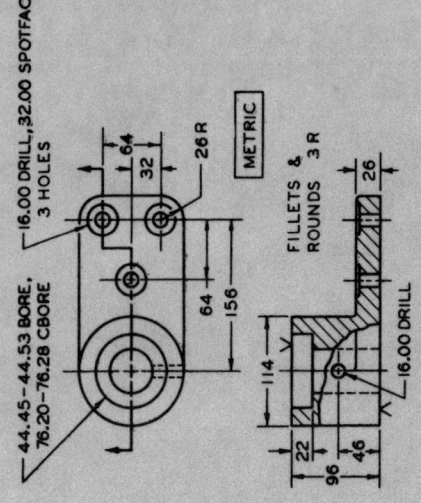

GUIDE BRACKET
C1 – 2 REQD
SCALE: HALF SIZE

Draw the two views mechanically.
Include all dimensions and notes.

BOTTOM OF DRAWING

LAYOUT WITH DIMENSIONS	DRAWN BY	FILE NO.	DRAWING
MECHANICAL DRAWING			12

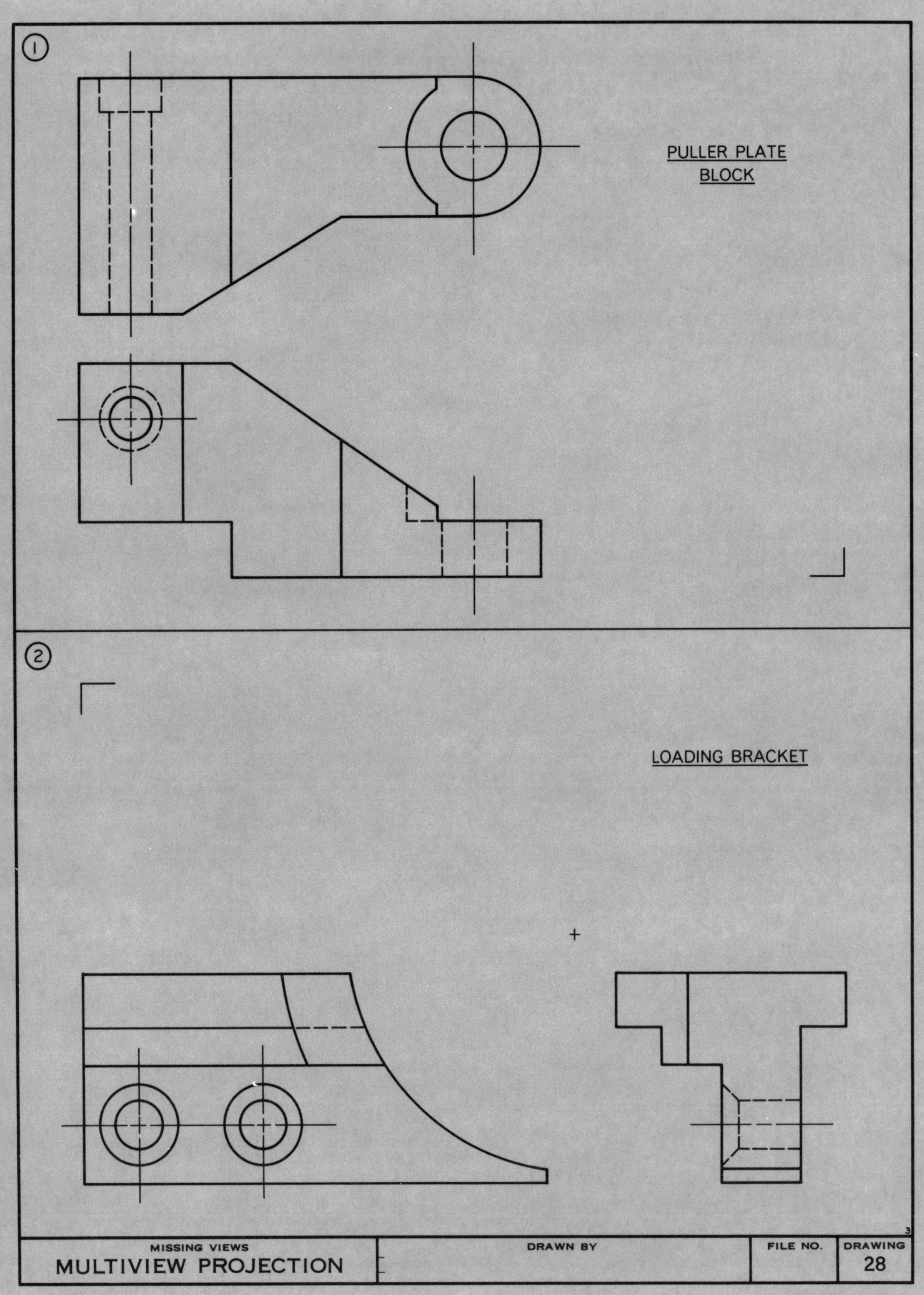

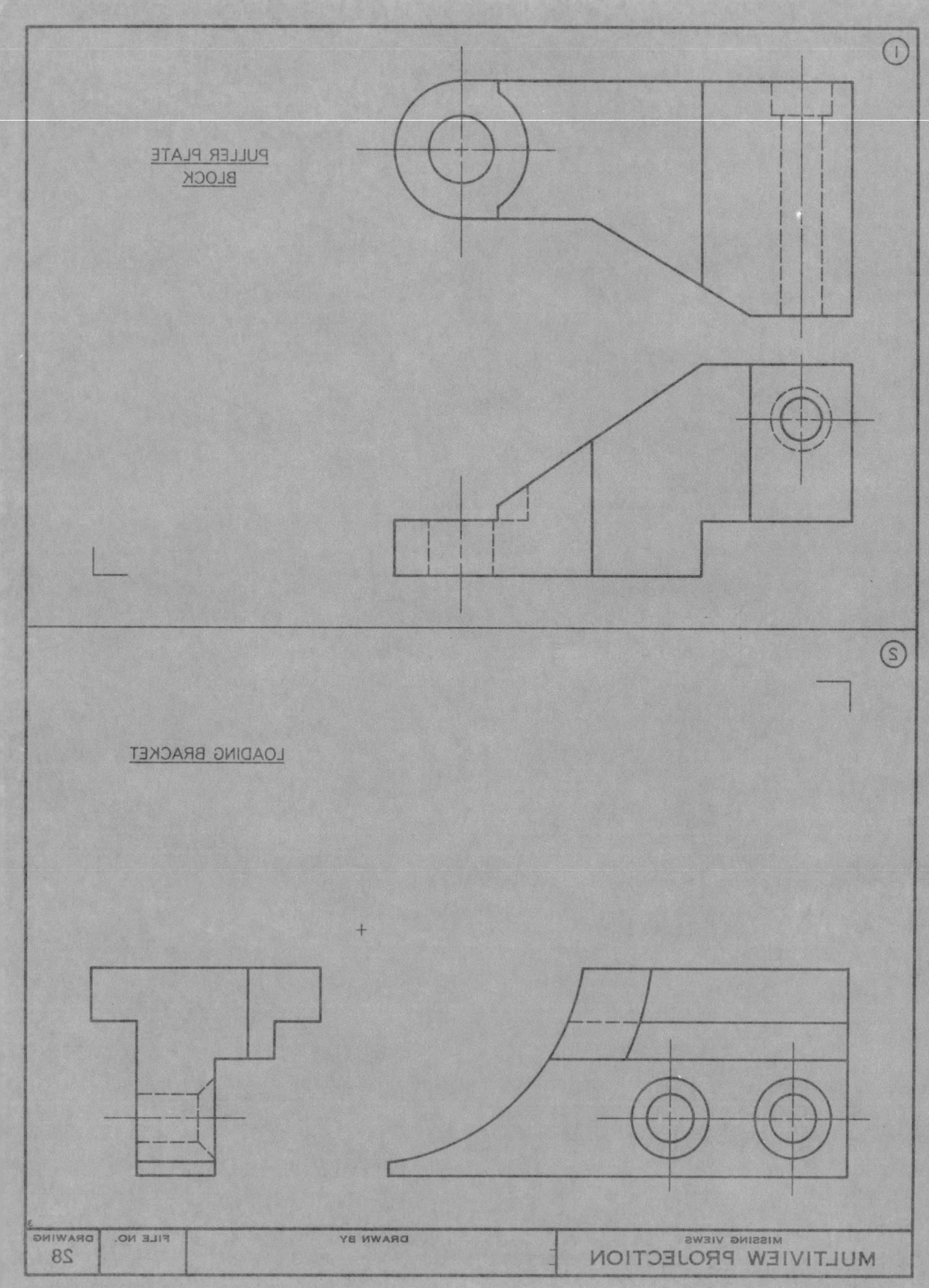

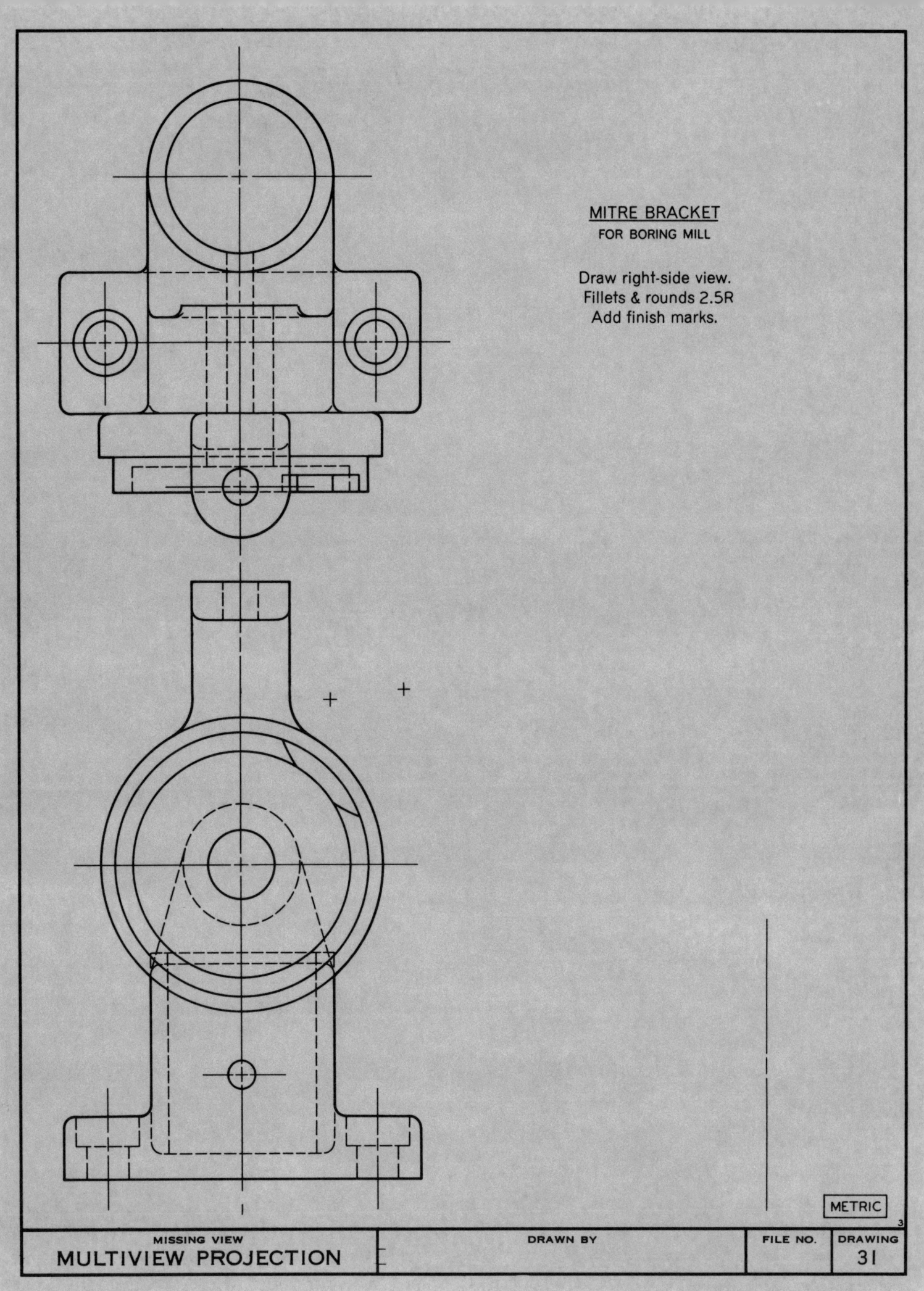

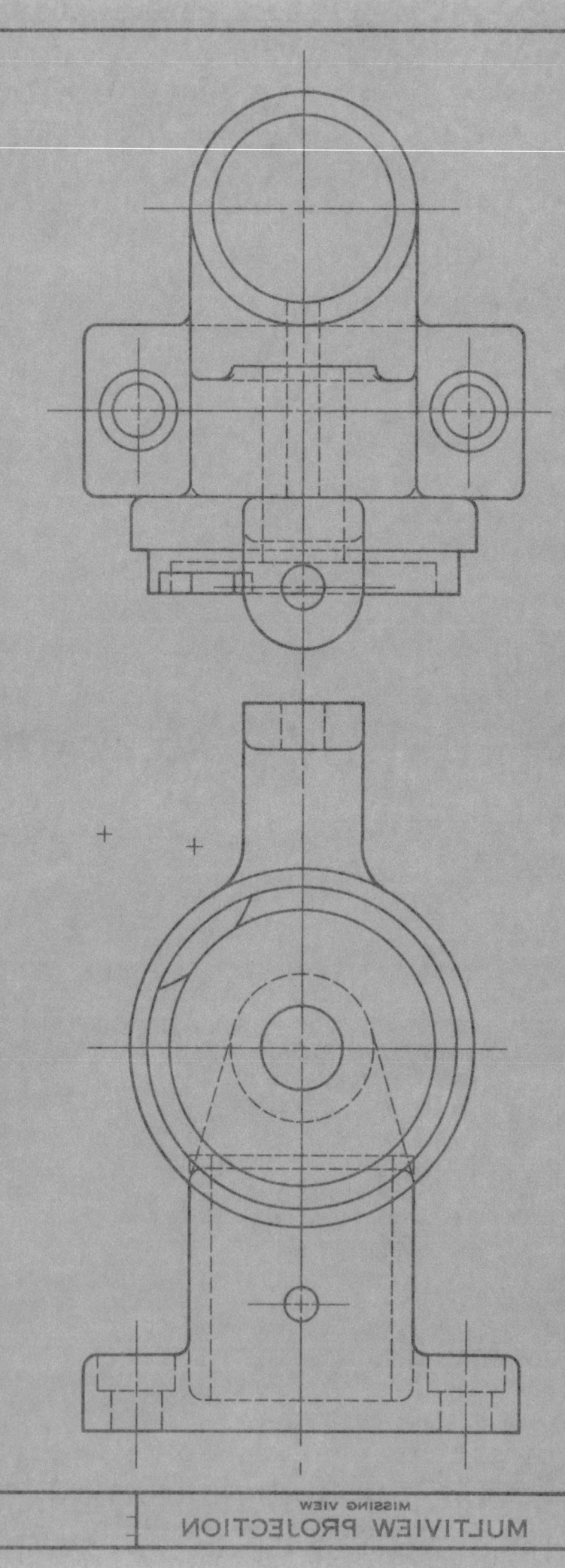

MITRE BRACKET
FOR BORING MILL

Draw right-side view.
Fillets & rounds 2.5R
Add finish marks.

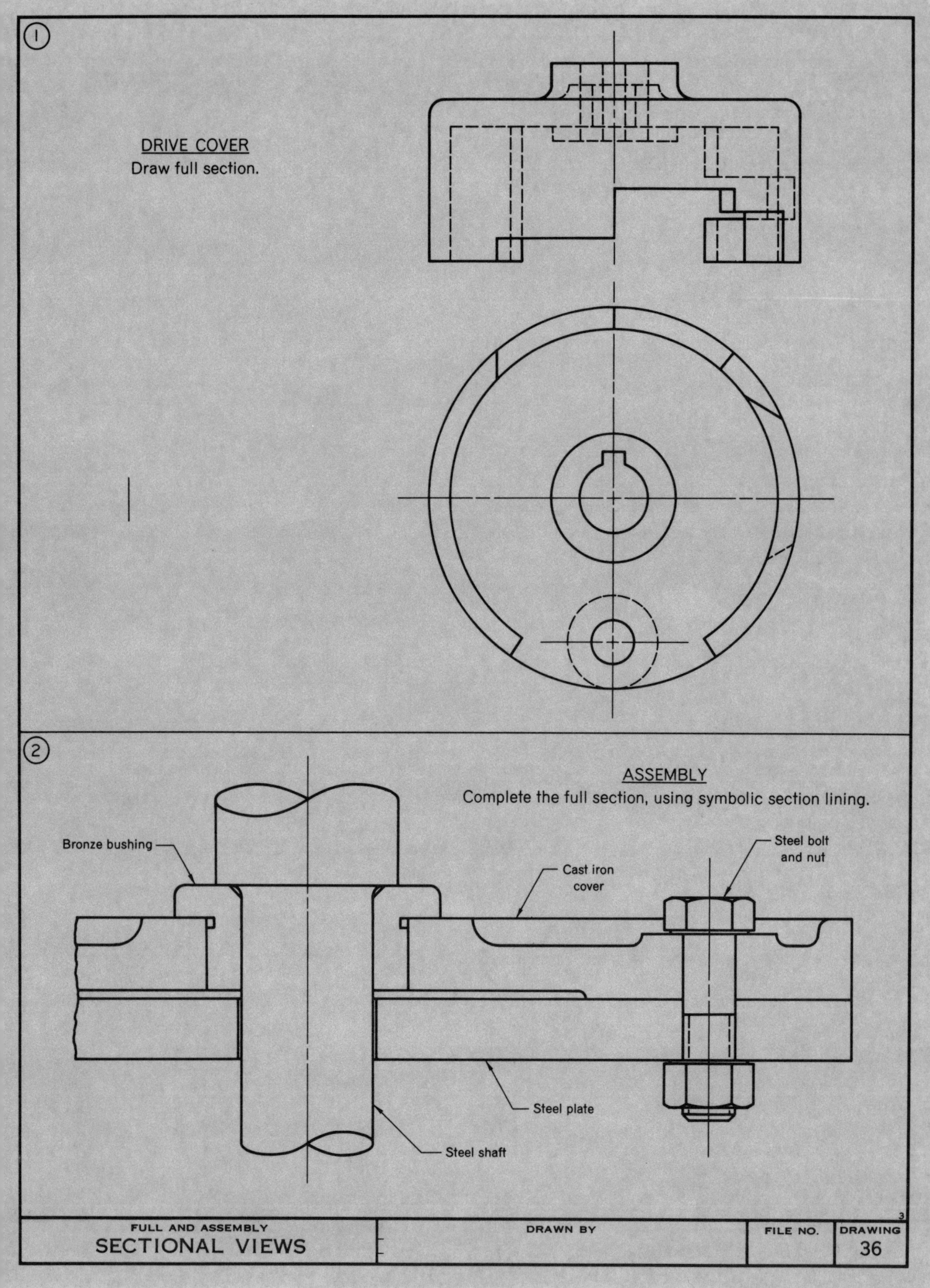

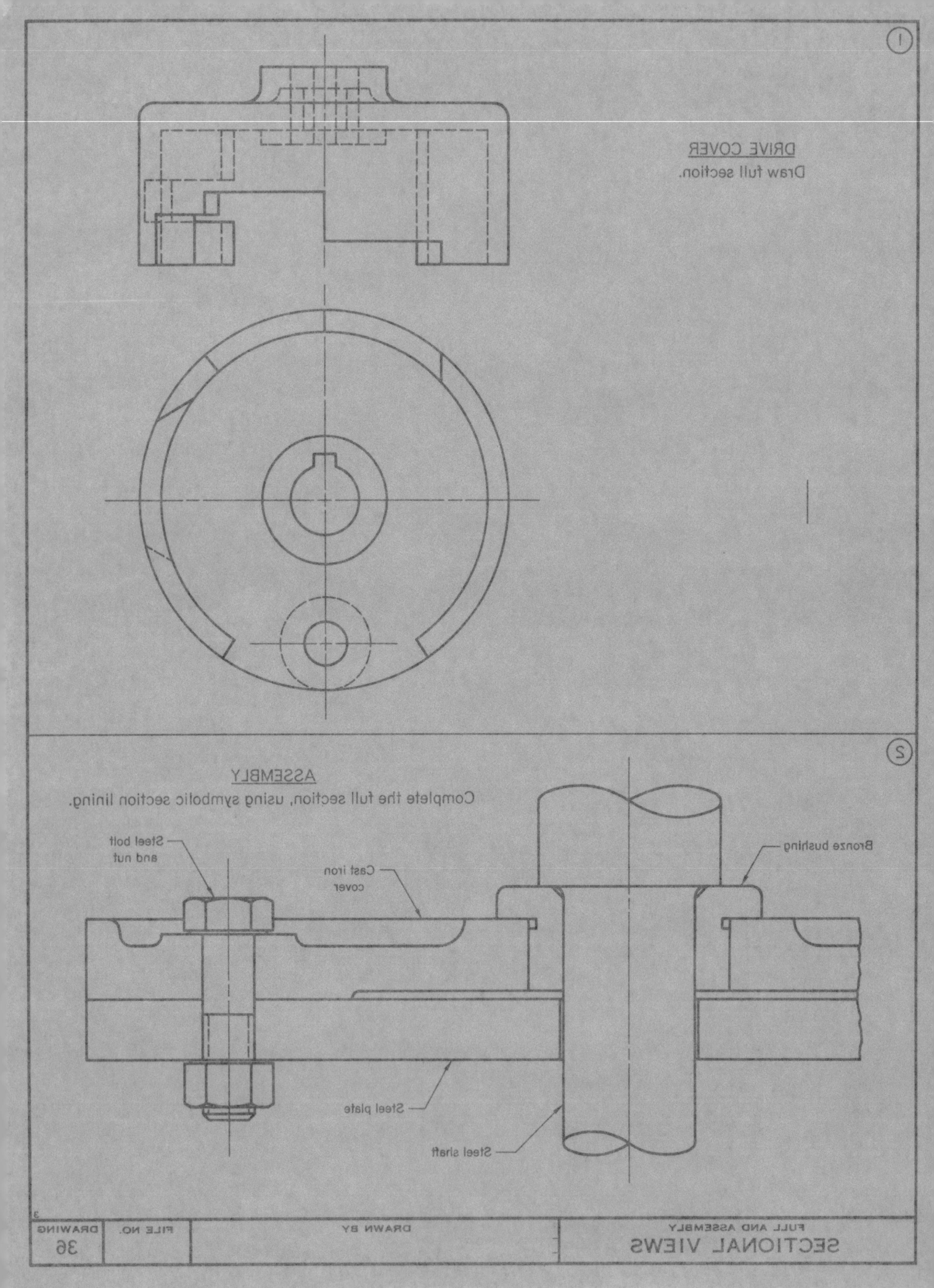

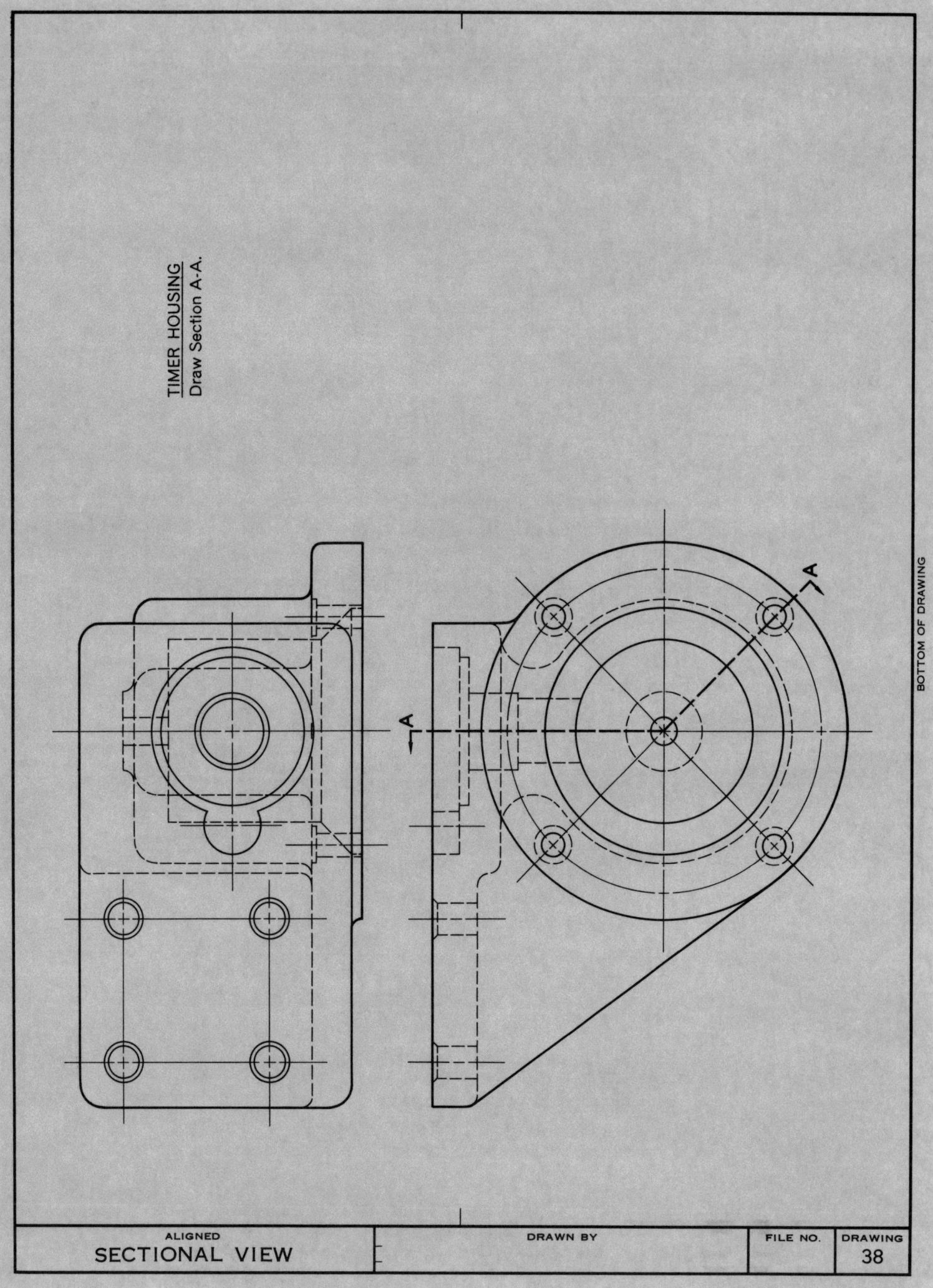

TIMER HOUSING
Draw Section A-A.

ALIGNED SECTIONAL VIEW

DRAWING 38

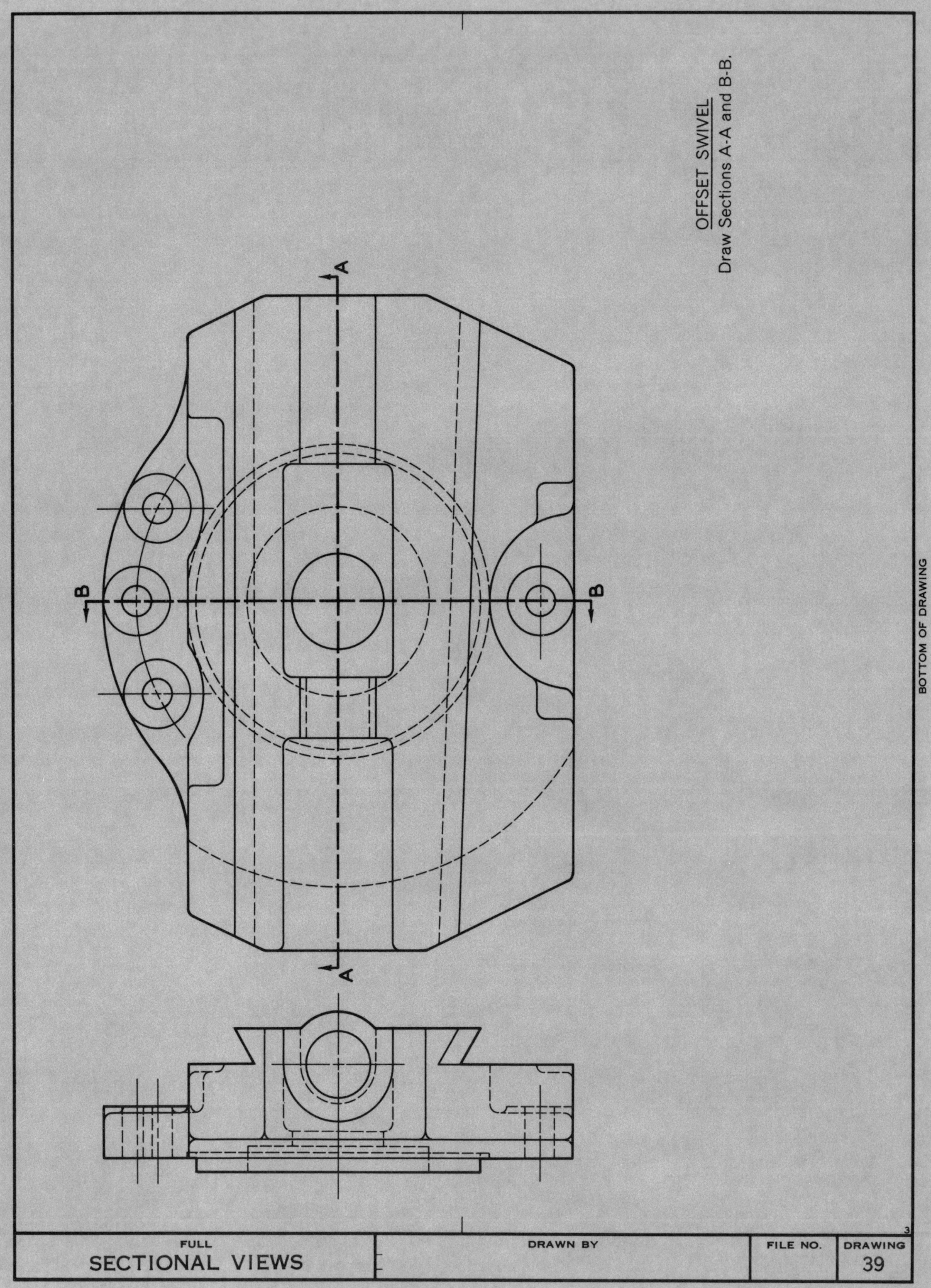

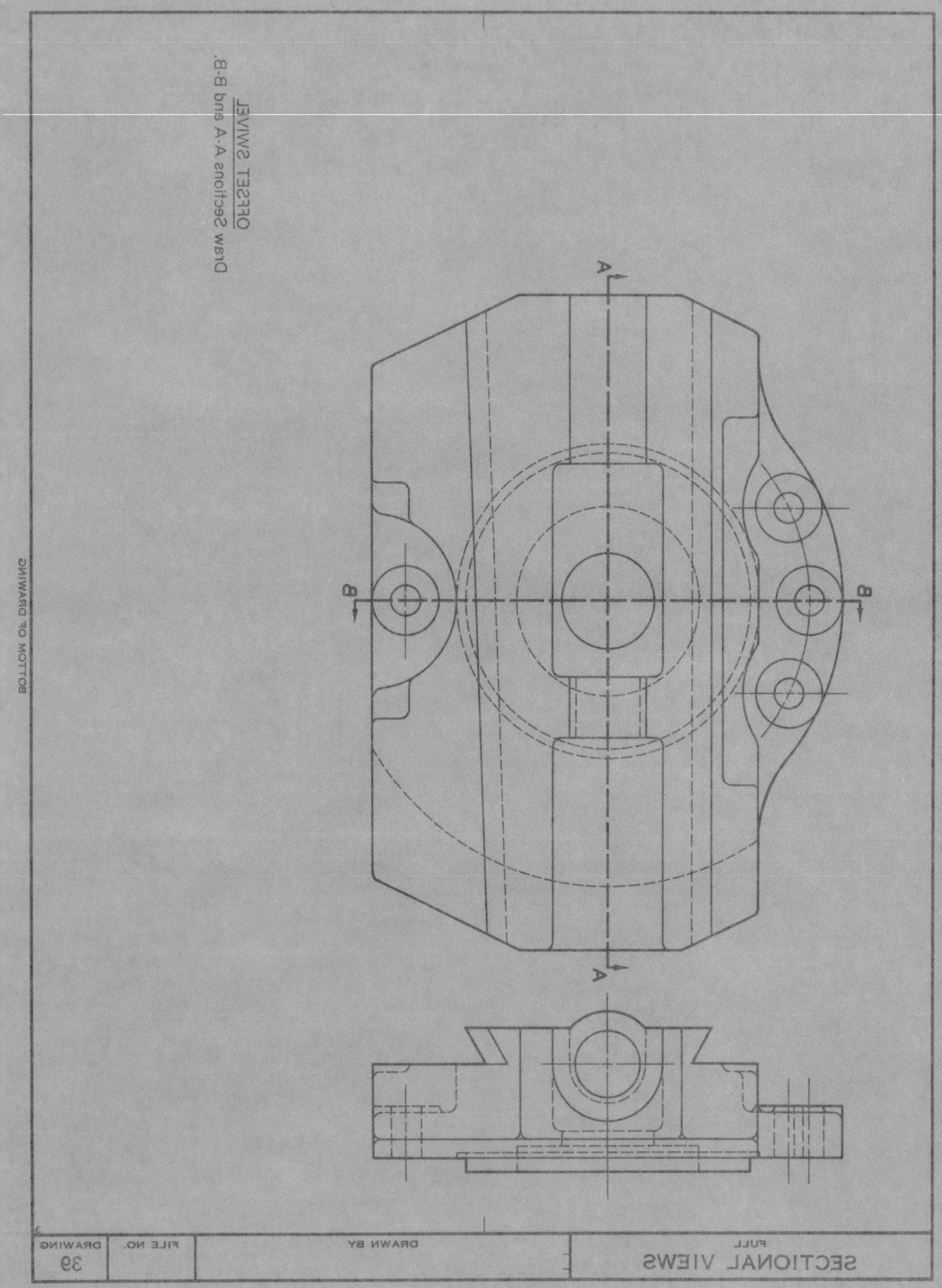

TOOL HOLDER BODY

Draw complete auxiliary view with direction of sight A, and partial auxiliary view with direction of sight B, each 18mm from front view.

Show all hidden lines.
Fillets & rounds 3R

Break line

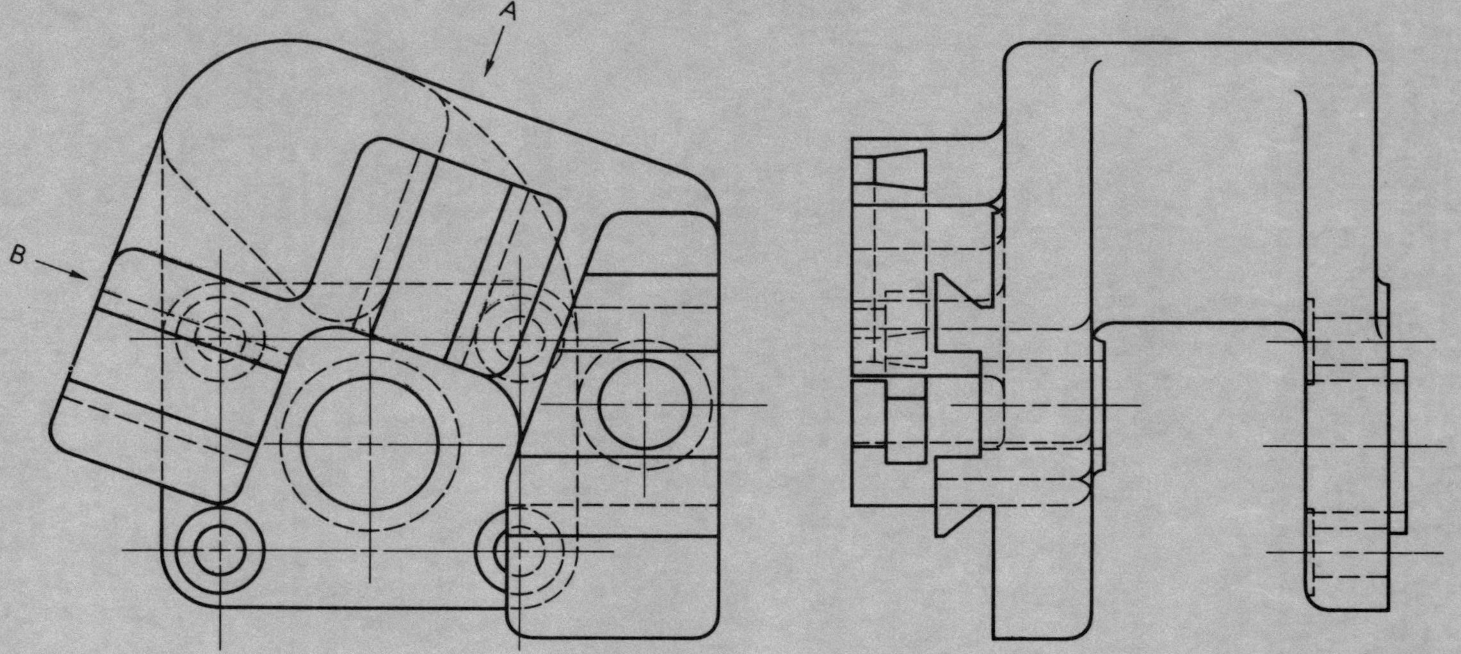

METRIC

BOTTOM OF DRAWING

PRIMARY
AUXILIARY VIEWS

DRAWN BY

FILE NO.

DRAWING 45

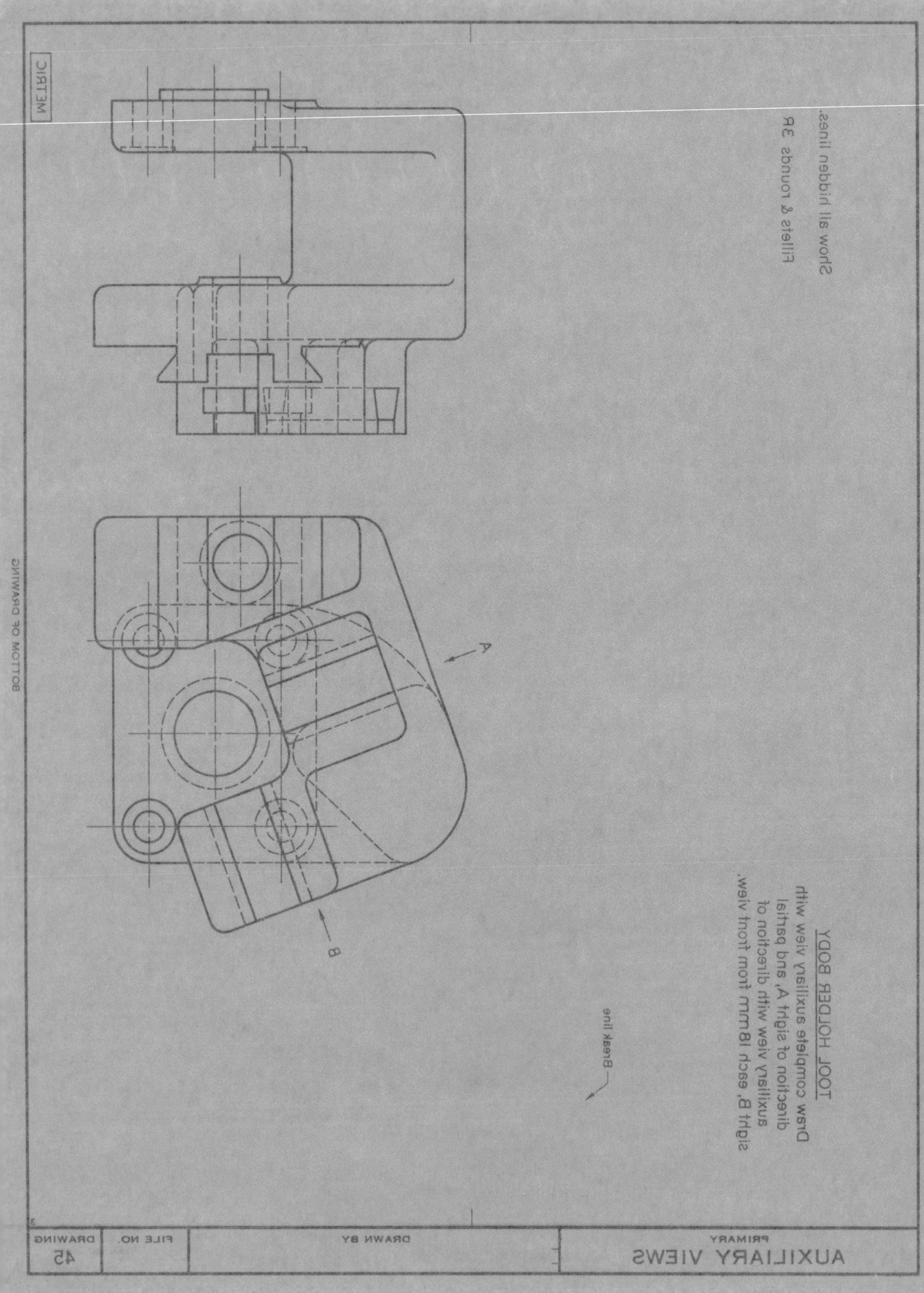

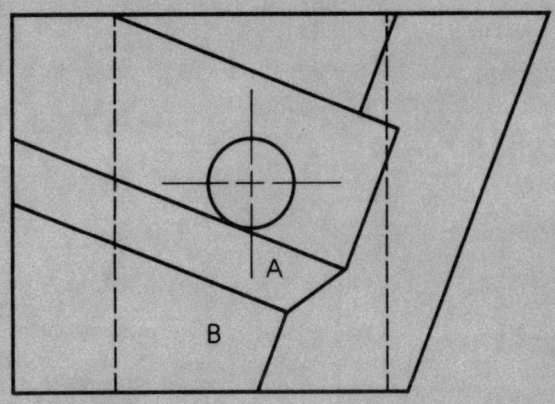

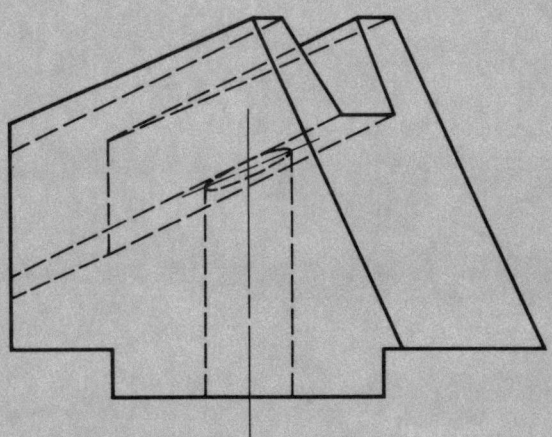

SLOTTED GUIDE
Draw complete primary and secondary auxiliary views to show end view of inclined slot. Include all hidden lines. Dimension angle between surfaces A and B.

| SECONDARY AUXILIARY VIEWS | DRAWN BY | FILE NO. | DRAWING 48 |

SLOTTED GUIDE

Draw complete primary and secondary auxiliary views to show end view of inclined slot. Include all hidden lines. Dimension angle between surfaces A and B.

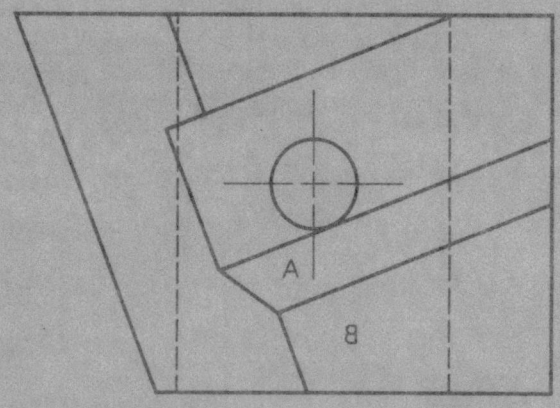

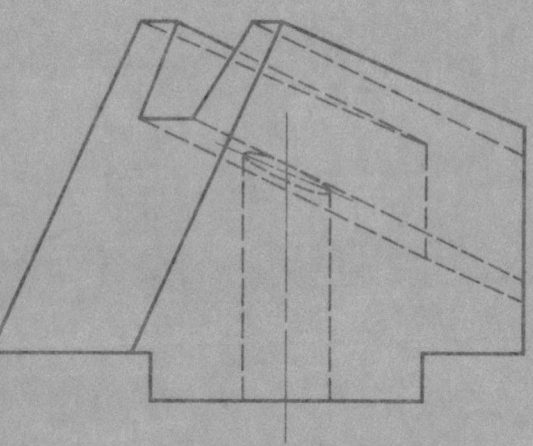

AUXILIARY VIEWS SECONDARY

DRAWING 48

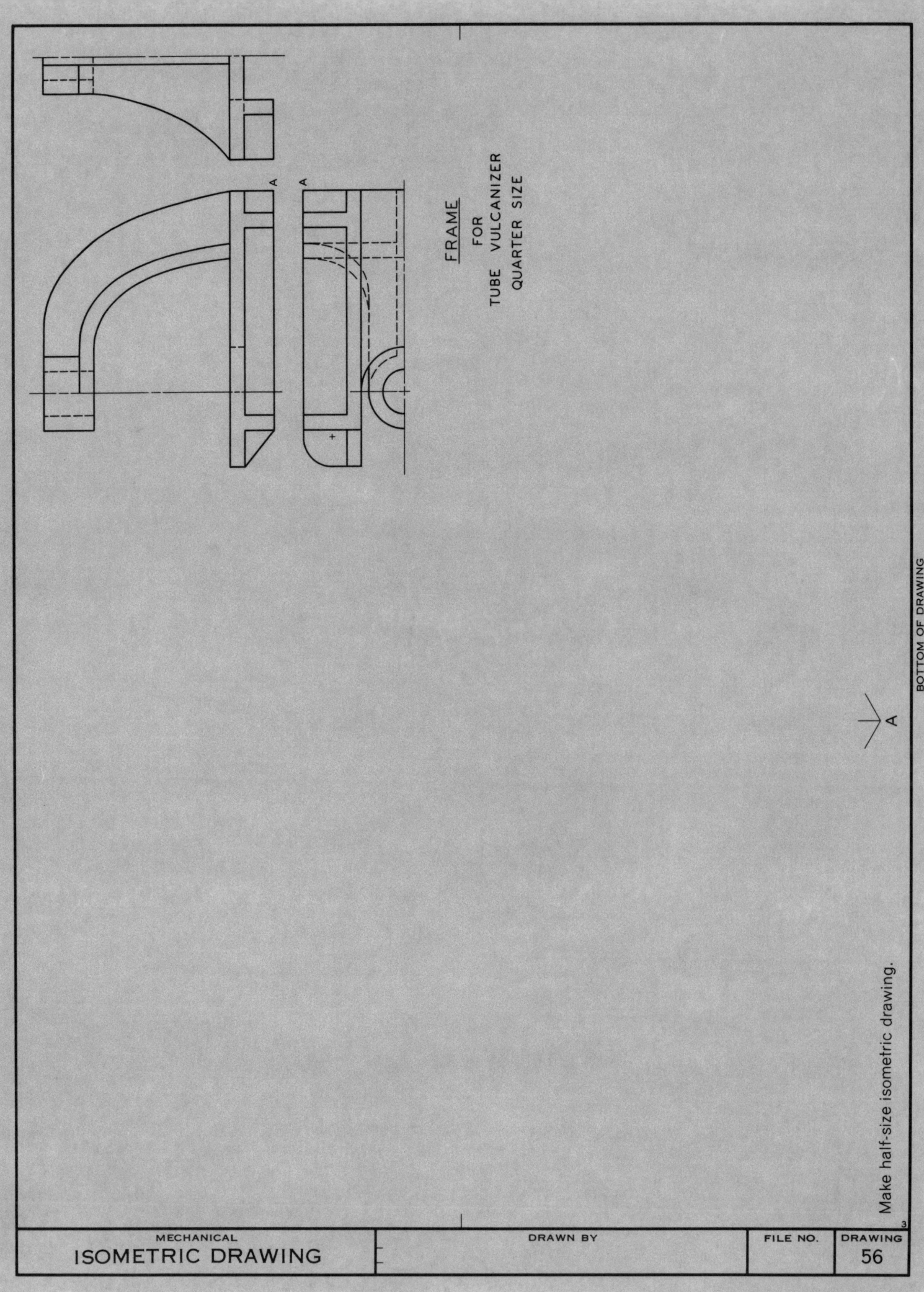

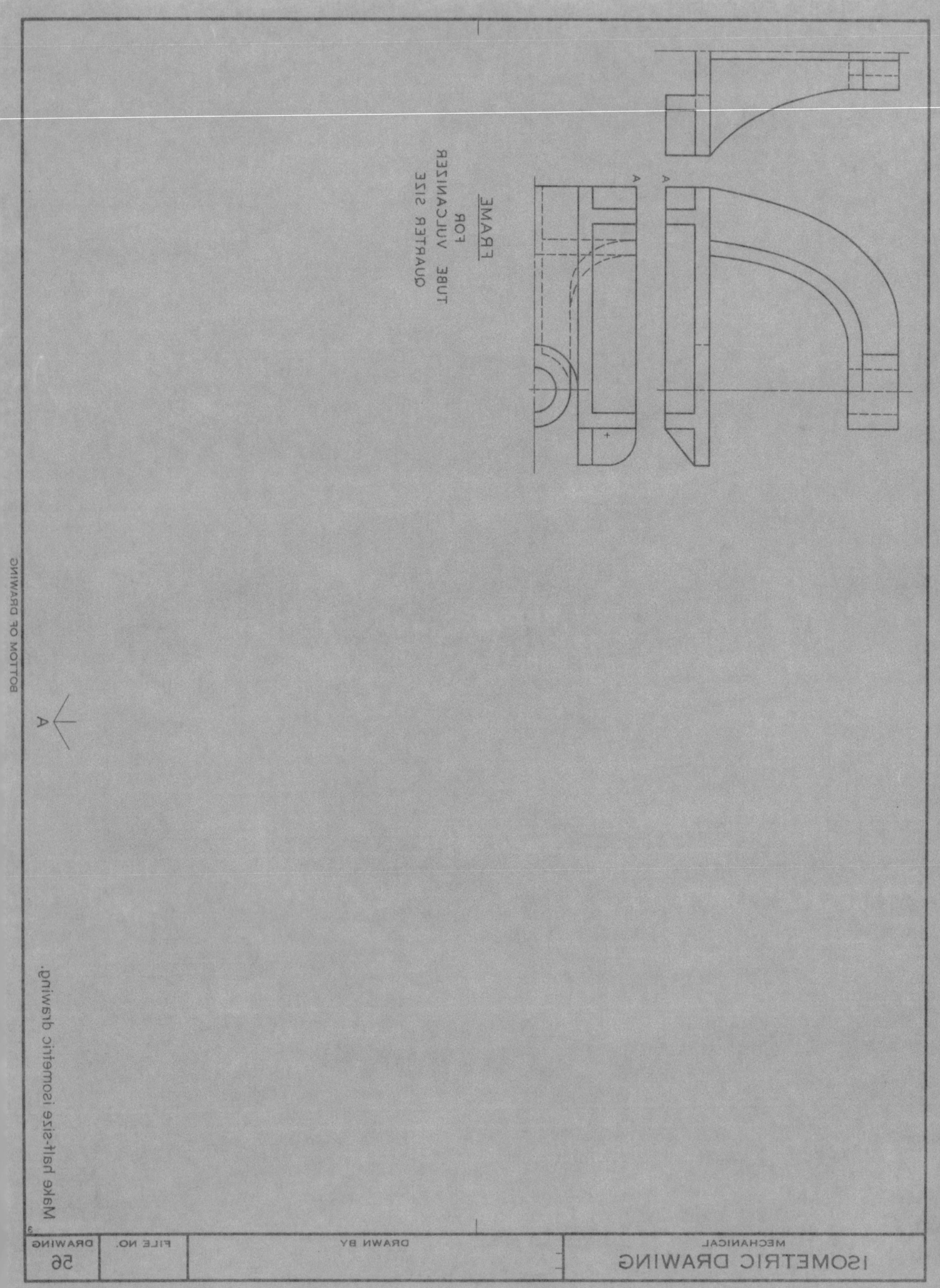

HOLDER
FOR THREAD ROLL
CRS – 1 REQD
FULL SIZE

Add dimensions mechanically.
Use complete decimal system.

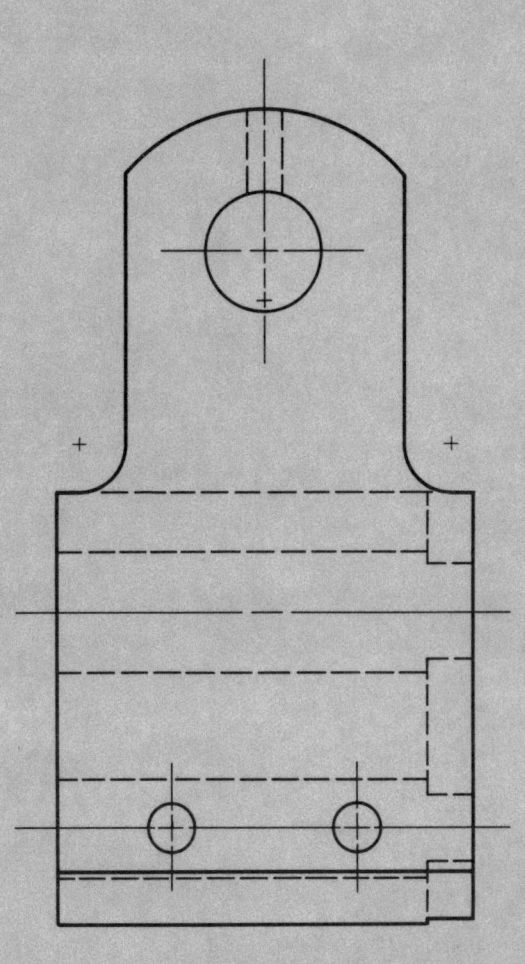
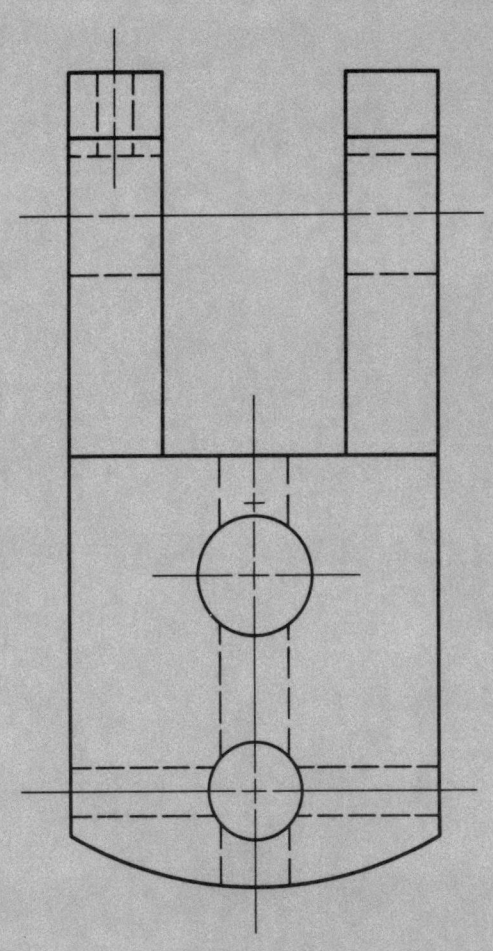

MECHANICAL DIMENSIONING

DRAWING 62

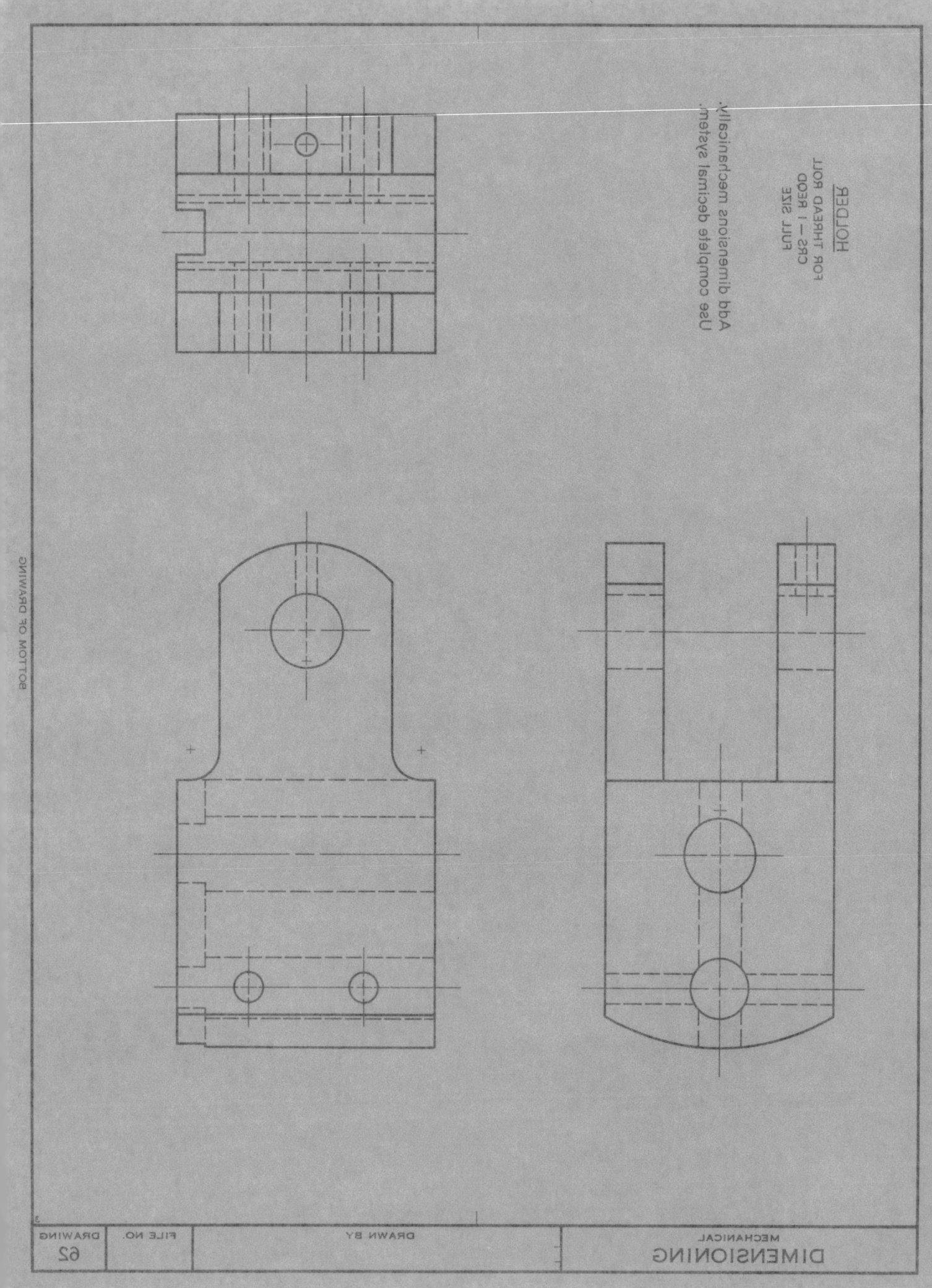

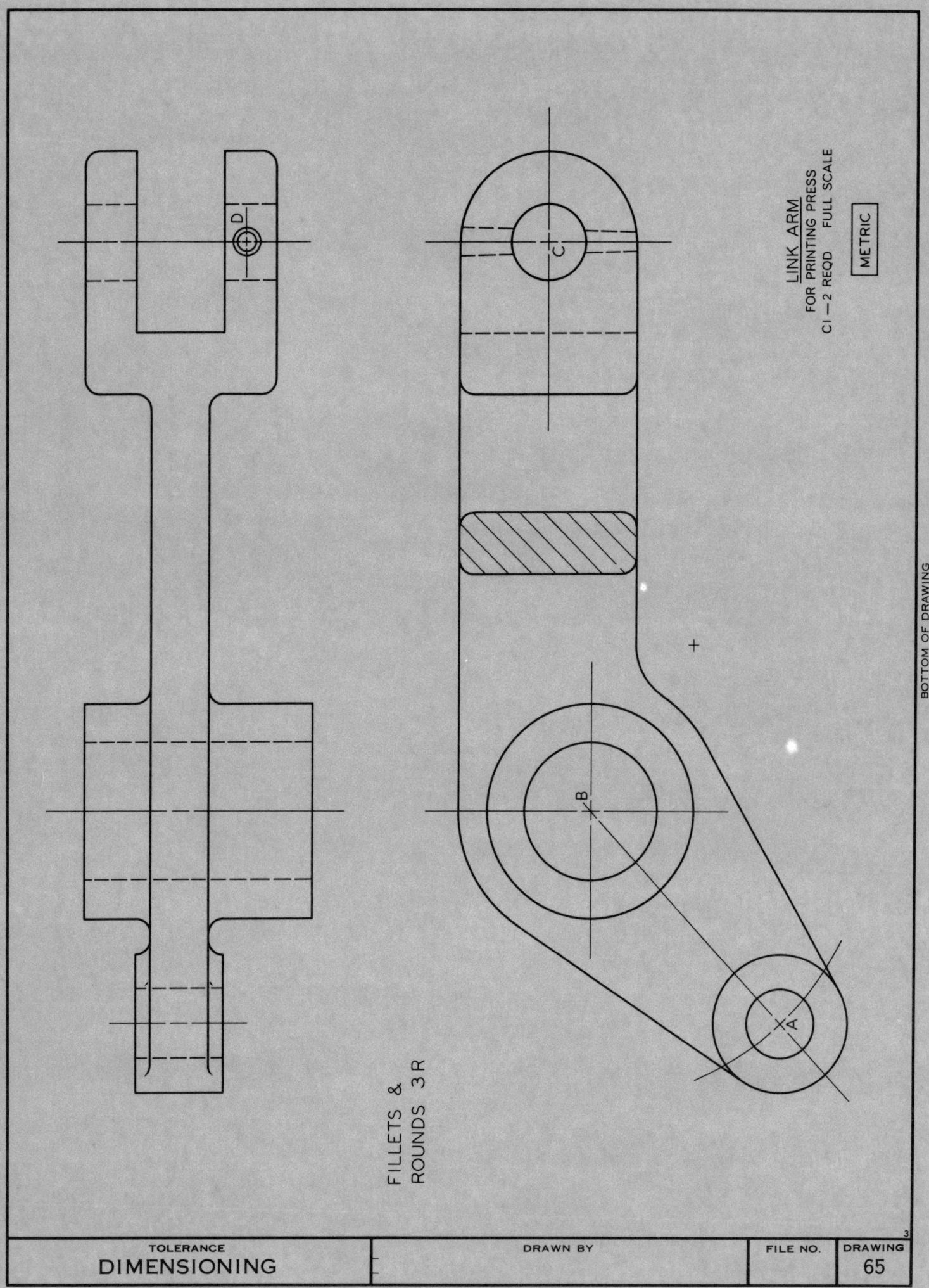

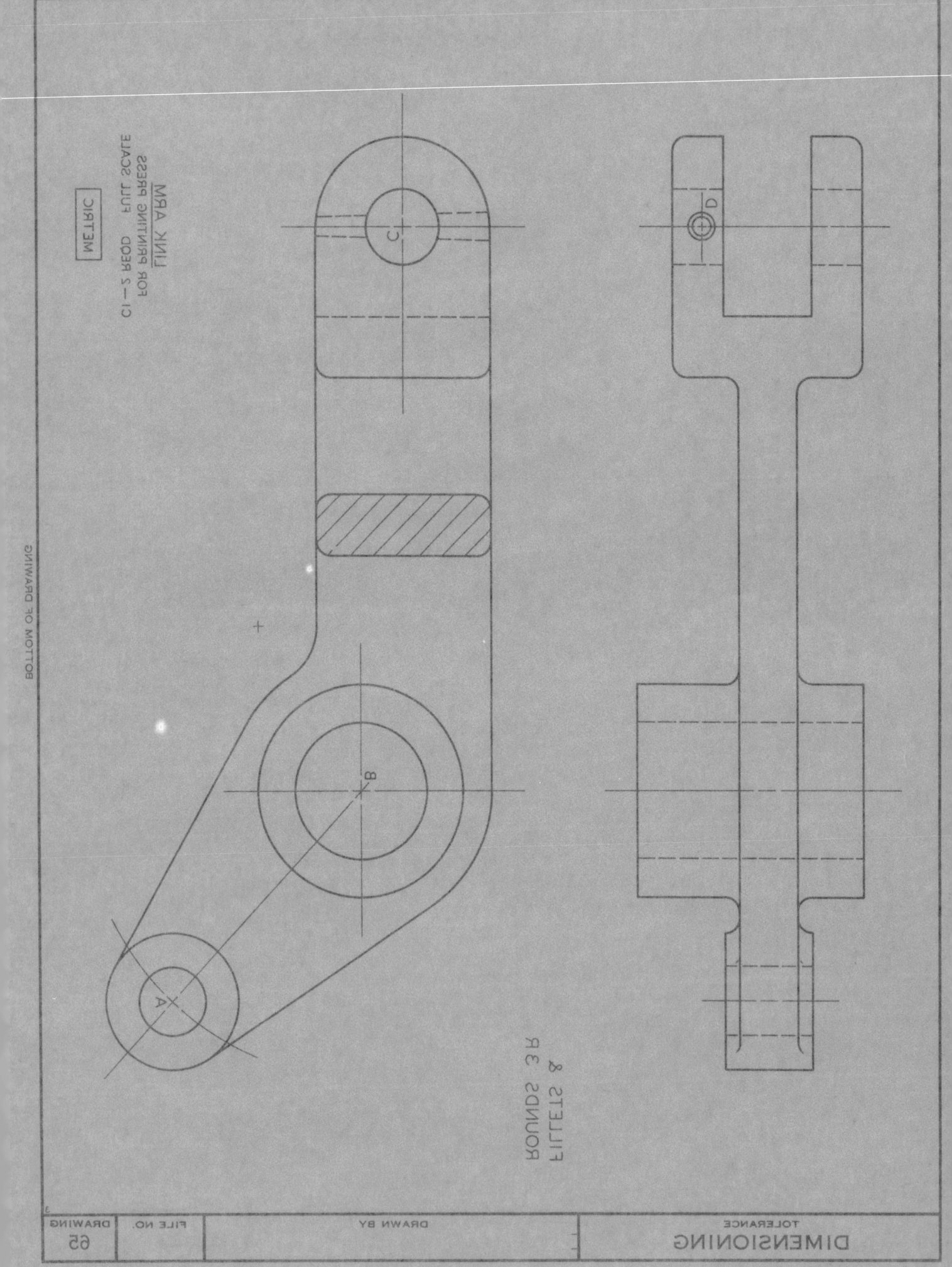

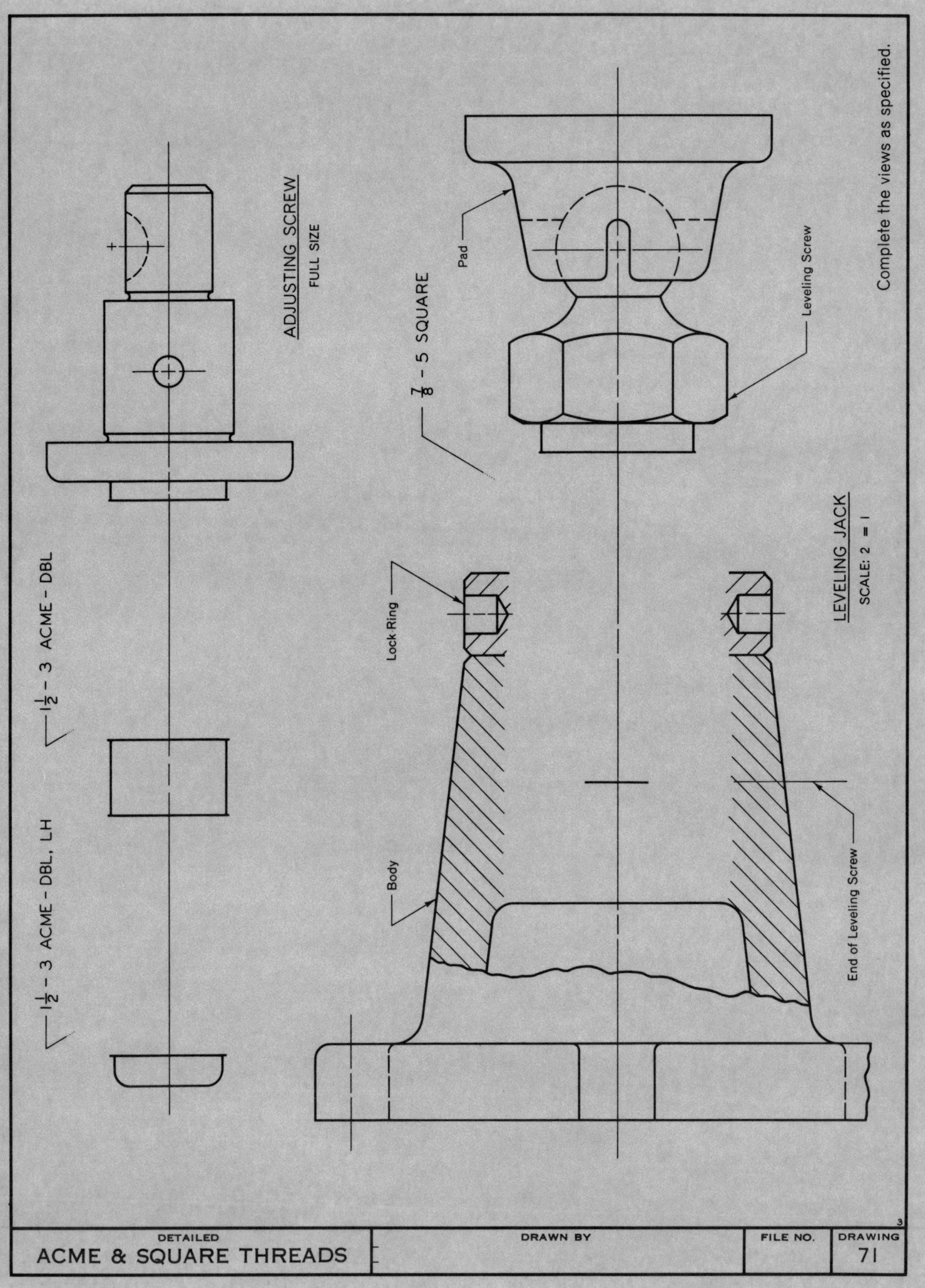

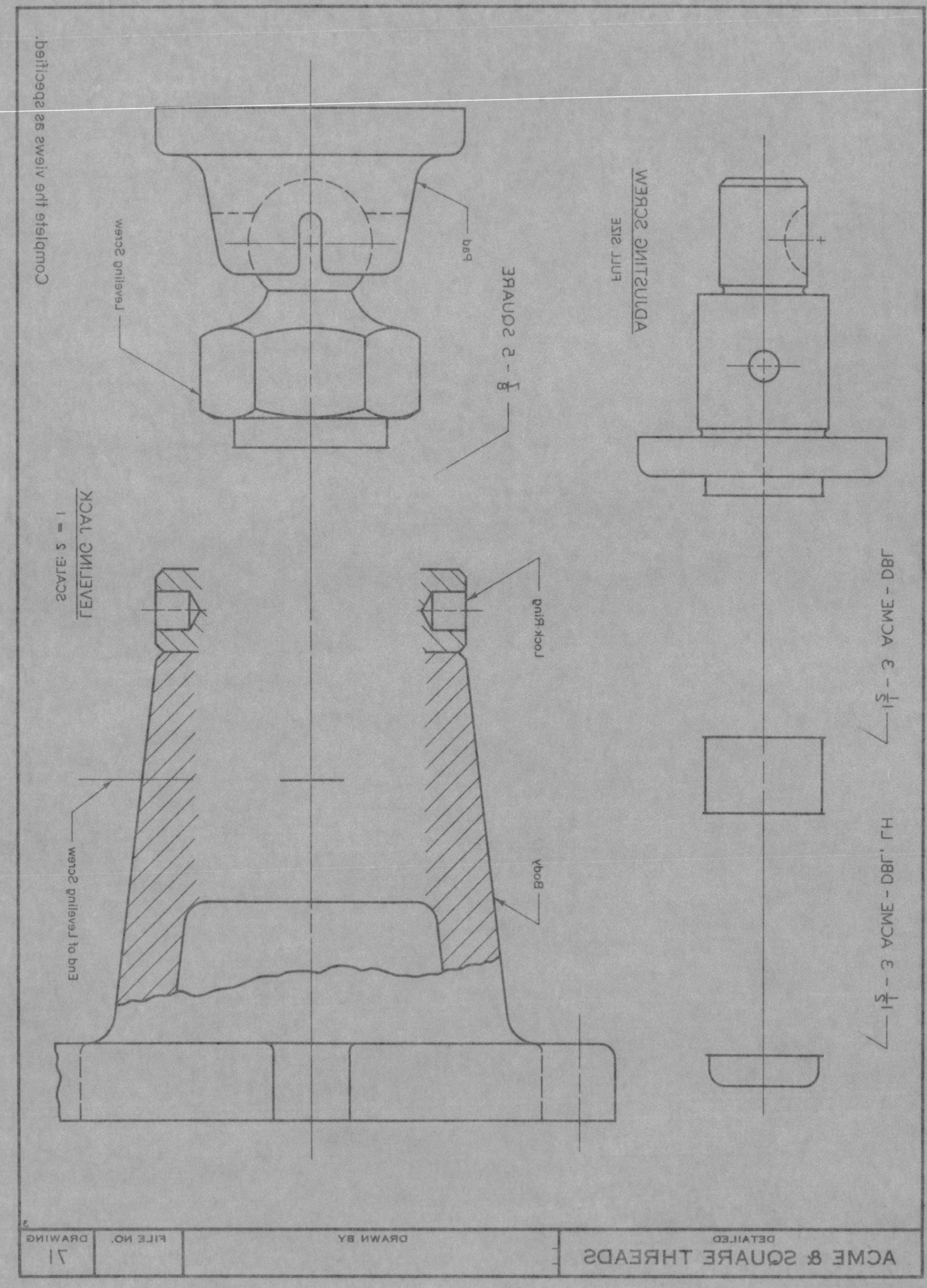

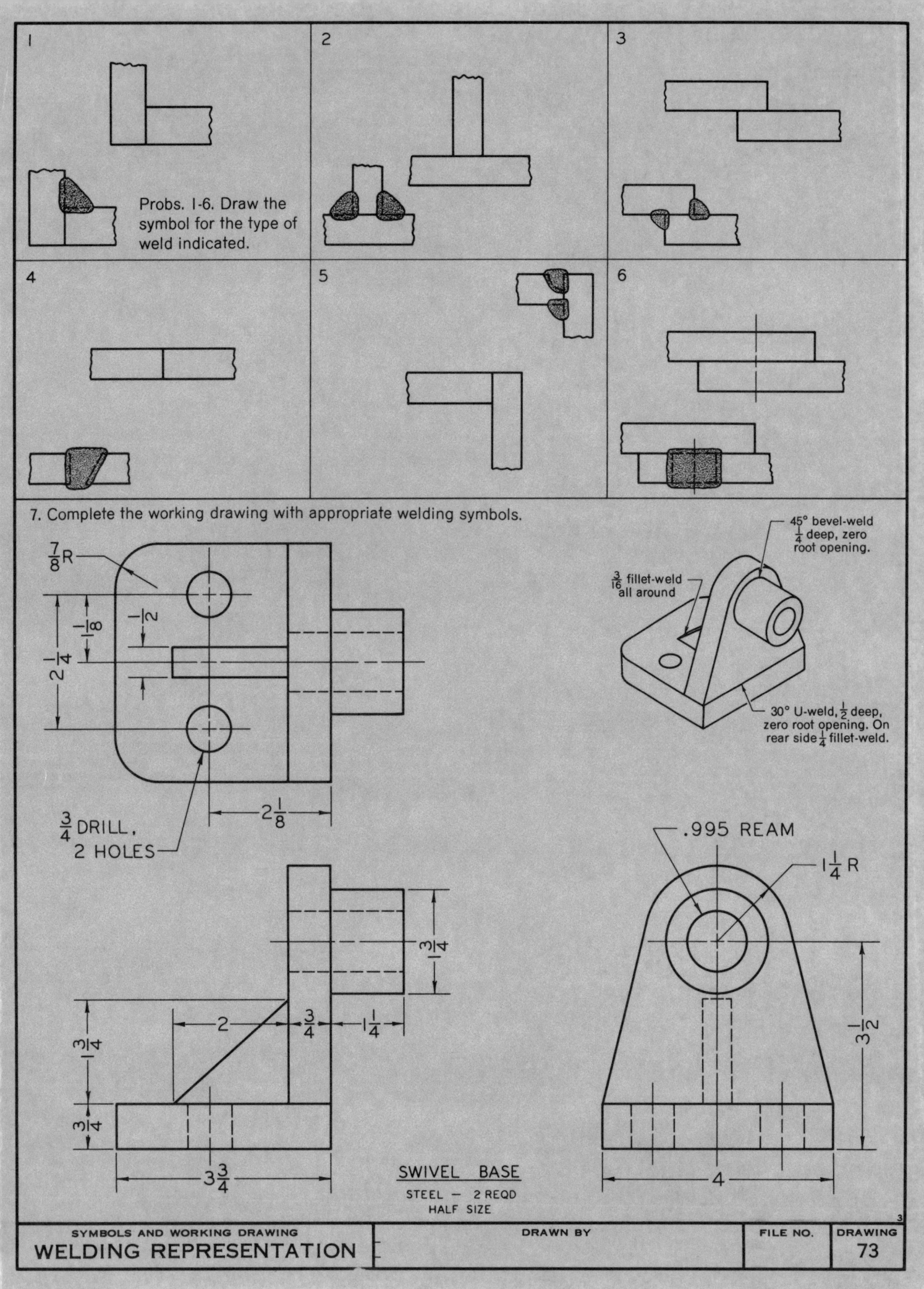

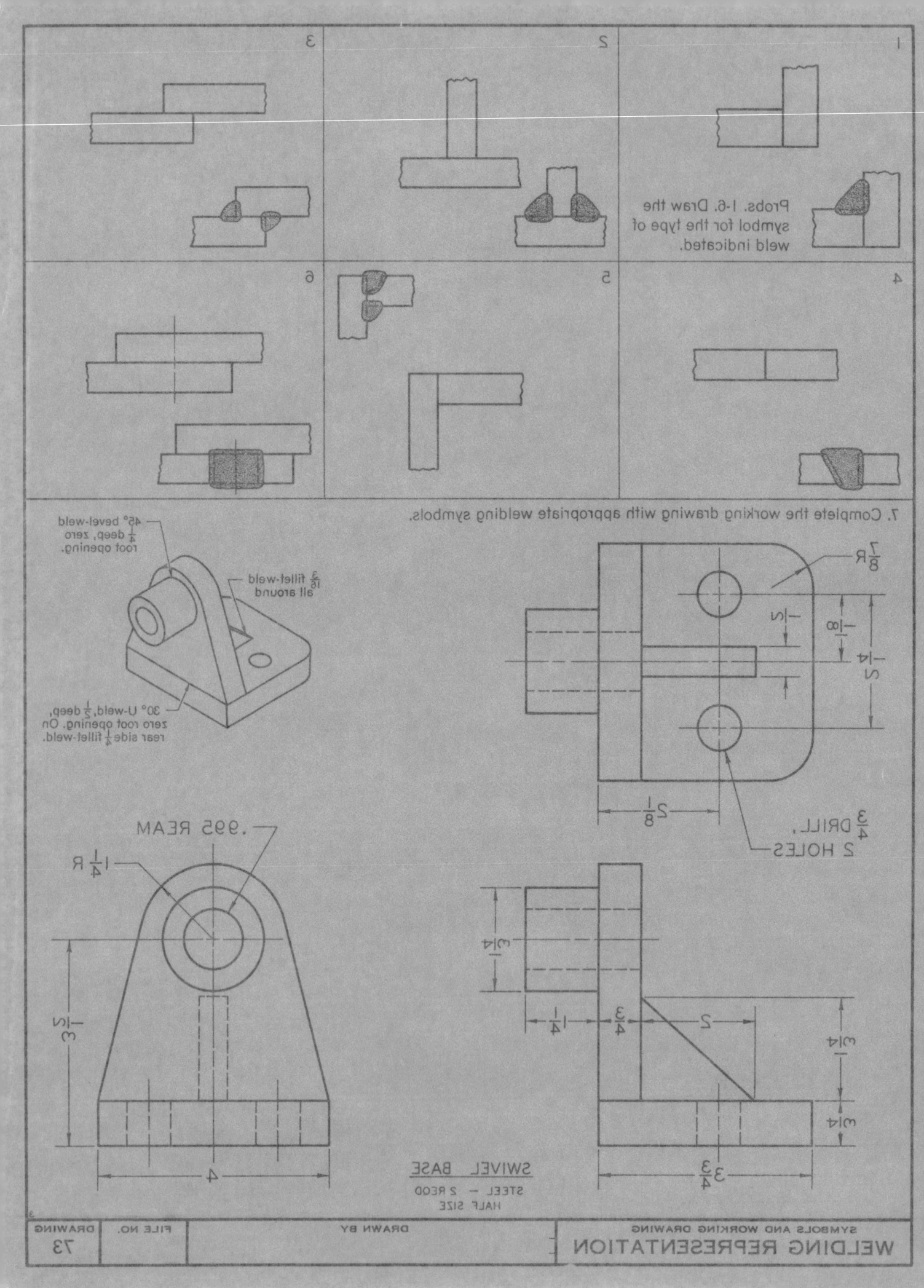